AF480791

Praise

This book shows that space is not distant science fiction; it is infrastructure, economy, and civilization engineering happening now.
Alex Melen | Co-Founder, SmartSites

As someone who flies every week, I've spent plenty of time staring out airplane windows wondering how humans went from dreams to balloons to rockets. *Everything Is Flying* connects those dots in a clear and engaging way.
Brandon Blewett | Author, *How to Avoid Strangers on Airplanes*

The imagery of space as an active environment full of systems, materials, and movement was inspiring. It makes you rethink our relationship with Earth and the cosmos.
Shawn Johal | Business Growth Coach, Elevation Leaders, Bestselling Author of *The Happy Leader*

The historical journey from ancient myths to modern orbital mechanics was beautifully done. Seeing how human curiosity evolved into engineering mastery was the most rewarding part of the book.
Casel Burnett | Vice President, LODI, and International Bestselling Author of *No Regrets*

This book made space feel personal and real. I walked away understanding how everything around us is constantly moving yet feels stable. Truly eye-opening.
Tamara Nall | CEO & Founder, The Leading Niche

Accessible, warm, and brilliantly clear, Ross's style makes space science feel welcoming rather than intimidating. This book proves that complex ideas can be both rigorous and deeply approachable.
Aaron Poynton | Bestselling Author, *Think Like A Black Sheep*

The book helped me see how space technology affects my daily life through navigation, communication, and weather prediction.
Trissa Tismal-Capili | USA Today and Wall Street Journal Bestselling Author

Mr Hamilton's book is a fantastic read for me as a sci-fi buff and a historian of science & technology in the Universe. The book reinforces so many unique scientific thoughts from my readings of Eric von Daniken 'Chariots of the Gods to the Mysteries of the Awesome Universe.' I and his book were inseparable for 3 weeks. An excellent work of science knowledge!
Darius Ross | Managing Director, D Alexander Ross Real Estate Capital Partners Interest LLC

EVERYTHING IS FLYING

From Ancient Skies to Deep Space—How Space Technology Shapes Life on Earth

BY ROSS HAMILTON

ISBN **979-8-99366-830-7** (hcv)
ISBN **979-8-99366-832-1** (ebook)

Library of Congress Control Number: **2026902371**

Dedication

For every explorer of ideas, bold and grand,
For those who forged the laws that help us understand—
For Galileo's gaze and Newton's great laws,
For Bernoulli's flows and Einstein's deep cause—
For rockets that climb past the sky's wide arc,
For probes and rovers exploring the dark—
For engineers' grit that blazes trails afar,
For my dad, my hero, friend, and North Star.

Acknowledgements

This book has been hanging around in my head for some years now, but it only came into being because of the love, support, encouragement, and patience of many people, and I am deeply grateful to everyone who has helped me along the way.

My first thanks go to Geeta Sidhu-Robb, whose encouragement pushed me to take the plunge, who generously shared her own experiences of writing, and who introduced me to the editing team, Leaders Brands. Mark Birch also played an important role by showing me how he approached his own self-published work and what it meant to see a project through from start to finish.

The flow of the book was influenced by the work I am engaged in at Space Network, and I want to thank Andy Campbell for the discussions we had on the broader space industry, and for helping me stay focused on the readers I wanted to write this book for: those who are space curious, but not yet technical or engineers.

Several people were kind enough to read early versions of the manuscript and provide thoughtful, detailed feedback. I am especially grateful to Professor Patrick Harkness, Professor of Exploration Technology at the James Watt School of Engineering, University of Glasgow, for the incredibly thorough comments and tweaks from cover to cover during first versions of the manuscript. I know this was not a thesis, but I appreciated the rigor of his review, which sharpened some of my explanations. I deeply appreciate the time that Dr. Nicholas P. Ross, Chief Executive Officer of Niparo, and Derek Harris, Director of Business Development and Communications at Skyrora, took to wade through the book and provide positive feedback and encouragement to continue the journey. I also want to thank Dr. Sharon Lèmac-Vincere for her candor and critical thinking, which helped me rethink some of the approach I had taken to the content and Further Reading. I also have to give a shout out to Dr. Stephanie "Wendy" Shorter, Ph.D. who has a true gift in sharpening bold statements, and proof-reading at warp speed. And then there is David Mackay, the Scottish astronaut who testflew Virgin Galactic flight to space, who was also very generous with his time, reading the book and giving valuable feedback and validation, and of course, for his kind and thoughtful words in the foreword.

I also want to acknowledge the professors and lecturers from my time at the University of Glasgow. The lessons in thermodynamics, fluid mechanics, and aerodynamics clearly stayed with me, and the act of explaining those ideas in plain language felt like a full-circle return to the foundations of my career.

To the team at Leaders Brands, I want to acknowledge the guidance, hard work, and logistical support you provided as we went through the editing, copysetting, design, and marketing of the book. You helped me appreciate the level of polish needed to go from a manuscript to a final product. Relatedly, I want to acknowledge that, in researching and shaping this book, I made some use of artificial intelligence tools to help with background research, refining some passages of text, and building the companion website, everythingisflying.com. In the end, however, the book reflects my own way of explaining things; any mistakes that remain are mine alone and will simply guide the next steps in my own learning journey.

Finally, I want to sincerely thank my wonderful family from the bottom of my heart. To my amazing wife, Laura, whose endless love, support, and encouragement have truly meant the world to me. She constantly inspires (and reminds) me every day to keep learning and growing. To my kids, Skye and Gavin, who inspired me to write this book in an accessible way that encourages others to discover the amazing and diverse world of space. And to my parents and my brother, Fraser, whose unwavering support has been a steady source of strength throughout my life.

Sláinte!

Table of Contents

Foreword

In 1675, Sir Isaac Newton wrote that he attributed his foresight to his predecessors - only by "standing on the shoulders of giants" was he able to see further. For those involved in aerospace test and development, in pushing the boundaries of what is possible and expanding the envelope, Newton's words are just as true today. We build on the work of our predecessors whose tested and proven theories have provided us with a platform from which to reach a little further than we have gone before.

In *Everything is Flying*, Ross Hamilton describes how our understanding of space began with our curious ancient ancestors trying to make sense of the motions of the night sky, and how our path to deeper understanding has been progressively built on previous knowledge, punctuated by many incredible insights from the historic giants of science before and since Newton.

As with almost any human endeavor, the path to progress is rarely a straight line. Along the way there will be occasions when previous theories and assumptions subsequently prove to be incorrect. In flight test and in spaceflight, we live with that reality every day. We must never be blinkered; we must always be open to new arguments or new information that may not necessarily fit our current model. Part of the excitement and fascination with space is that our knowledge of it is still constantly changing.

One of the most important themes in this book is that Hamilton shows us, at each stage, how progress in space has benefited our lives on Earth. He clearly explains and demonstrates that the often considerable cost and difficulty of accessing and using space have resulted in an extraordinary range of gains for us back on the surface. It becomes clear that human challenges and subsequent endeavors are important not just in their primary aim, but in how technological advances and a deeper understanding in one field can reap huge rewards in others, even if the original aim is unfulfilled.

In my own career, I have seen how ambitious goals in aerospace lead to unexpected advances in safety, materials, communications, and our understanding of our planet. New challenges we set ourselves can lead to all sorts of unexpected benefits. The search for understanding and the desire to explore are primary human traits; these are exemplified by our exploration of space.

Everything is Flying is an easily accessible story that creates a concise mini-encyclopedia of space history, while demystifying the technology and the science. It is a very readable resource of facts and information. As an argument for continued exploration and endeavor - from someone who has been fortunate enough to see the Earth from the edge of space - I believe it deserves a place on everyone's bookshelf.

David Mackay
Former Chief Pilot, Virgin Galactic

Introduction

Everything is flying. Every world and star you can see, and every atom in your body belongs to an ancient and continuous system of motion–whether you sense it or not.

The Sun sweeps around the center of the Milky Way at about two hundred kilometers per second, carrying Earth and its planetary siblings with it. Earth spins on its axis and orbits the Sun; the Moon orbits Earth. Comets and asteroids follow elongated ellipses through the dark.

Our air never rests, driven by solar heating and turned by Earth's rotation. The tides rise and fall under the Moon's pull, with the Sun adding its steady tug. When charged particles from the solar wind stream along Earth's magnetic field, the sky fills with auroras–curtains of light against space's dark.

Above our heads, satellites and space stations flash across the sky, orbiting many times faster than a jet. Rockets roar and accelerate off the planet, speeding into orbit and launching explorers and machines into the unknown.

And within this grand whirl, we cling to our wee, warm, blue world while everything rushes on at breathtaking speed. Only through science and the language of engineering do we map, measure, and understand our place in this chaos; only by building machines that can also learn to fly do we gain insight into how the universe truly works.

Ultimately, we are but tiny passengers in this wild, vast web of motion that binds us to the planets, stars, and all living things. The deeper we look, the more we realize: everything is flying, and so are we.

The Sky Was Never the Limit

Look up at the night sky. That vast, star-filled canvas has been humanity's constant companion—our guide through uncharted waters, our way of life, our source of stories, and a spark for our imagination.

What was your first experience of space? For me, growing up in the 1970s, it was *Star Wars*. The excitement of the Millenium Falcon (best spaceship ever), hyperspace jumps, Darth Vader, and the Jedi—it all blew me away. It was thrilling to think of other planets filled with funny creatures and brave humanoids.

That early wonder led me from building paper and LEGO models of X-Wings to assembling plastic model aircraft kits from Airfix. I still remember the sticky glue, messy paints, and fragile decals as I raced through the detailed instructions so I could hang it from my bedroom ceiling and move onto the next one.

And while many people got to experience the Moon landing on their TVs around the world, for me it was the launch of the Space Shuttle program that has been burned into my brain. Aside from the marvel of an airplane-looking thing impossibly hanging off the side of a rocket on one side, it was what it did in subsequent missions that was just incredible: spacewalks, robotic arms, and building the International Space Station (ISS).

A few years later, I found myself at Rolls-Royce Aero Engines which sponsored me through my Aeronautical Engineering degree at the University of Glasgow. That first summer was my first experience of being on a proper factory floor and it was truly unforgettable: the smell of coolant as massive lathes carved aircraft engine parts, the intense heat as molten metal was poured into turbine blade molds. We learned how important quality control was in manufacturing very complex engines, while also sometimes needing to wield a "Friggin' Big Hammer" (or the equivalent saying) to get things working.

Over the next four years, I stripped down and rebuilt engines, working with teams of engineers, craftsmen, welders, electricians, and others much older than me. You had to learn about team dynamics and culture. Academically, I got immersed in the depths of aerodynamics, fluid mechanics, thermodynamics, and aero structures. I got to design aircraft on drawing boards with pencils and rulers and build mathematical models of helicopters flying in turbulence.

Even as my career journey led me through various fields like finance, technology, advising startups, and founding companies, the core lessons from those early days have always stayed with me. At the heart of it all, I always believe it's about the people– their dedication, pride, and shared commitment to something bigger than ourselves.

As life sometimes does, I found myself completing the circle with a new adventure: becoming a founding executive at the Space Network, a Scottish company focused on commercial space ventures. You might wonder, what does that really mean? For me, it's about helping to create innovative space solutions that attract investment, customers, partners, suppliers, and talented people. And once again, it's all about the people. Just like any other business, space businesses are just like regular businesses: they are team sports.

At the same time, there are global trends that are affecting our daily lives. Global geopolitics are changing, and space is becoming competitive in everything from internet access to national security. Governments and the private sector are investing billions in space, while the public still thinks it is just about astronauts and rich people messing around. Meanwhile, businesses and policymakers need a narrative to understand how space is the new global infrastructure that the world and economies depend on.

That is why I wrote this book. To open the door for everyone to learn about space–how space technology works, how it shapes our daily lives, and how you can be part of this journey. Whether you're an aspiring engineer, a scientist, interested in law or business, an educator, a teenager or student still figuring out your path, or simply "space curious," this book is for you. I want to help you understand a little more about this vast frontier that belongs to all of us.

Standing on the Shoulders of Giants

Every breakthrough in space is built on the vision and labor of those who came before. Throughout history, great men and women have explored and settled three great geographies: land, sea, and air. Now, we stand at the dawn of a fourth: space. Like the others, space is far more than a place to visit; it's a realm to explore, inhabit, understand, and ultimately call our own.

We've moved from stargazing to stepping foot on other planets, transforming those distant points of light into destinations within reach. Just as we adapted to life on the ocean and in the sky, we are now learning how to live in the vacuum of space. Technologies born from submarines and high-altitude research help shape space habitats. Spacecraft innovations now support life in Earth's harshest environments—from Antarctic research stations to desert solar farms.

Space is not just another industry; it's a new geography extending human civilization and sparking innovations that benefit all humanity. It is humanity's most ambitious problem-solving laboratory that often leads to solutions for Earth. It demands expertise from physics, chemistry, biology, economics, law, politics, and many other fields.

Water recycling systems developed for space are now improving conditions in refugee camps. Solar panel technology originally created for satellites is revolutionizing clean energy on Earth. Farming techniques meant for Mars help us grow food more efficiently here at home. The smartphone in your pocket connects through satellites orbiting overhead; every weather forecast relies on space-based observations; global financial markets use satellites to execute trades; GPS is now on your smart watch and phone. Innovations born in the space sector keep solving problems on Earth.

Behind these innovations is a legacy built by brilliant minds—astronauts and rocket scientists, philosophers, oceanographers, climate researchers, and engineers. We stand on their shoulders, seeing further because of the foundations they built. They understood that the challenges we face in space often reflect those on our own planet. Their work shows us that exploring space is not just about leaving Earth. It is about understanding Earth and ourselves better.

As we face unprecedented challenges—climate change, resource scarcity, and many others—this new geography offers more than escape. It offers hope, inspiration, and new ways to confront the impossible. The harshness of space forces us to invent technologies that are more efficient, sustainable, and resilient—the very qualities needed on Earth, now more than ever.

Space Curiosities

My hope is that this book addresses and sparks your curiosity about space—how it all works, where it came from, and where it's heading. You'll discover how space technology and Earth technology grow more intertwined every day, creating solutions that work both in orbit and here on Earth. I hope you glean some useful and fun facts along the way.

My goal is to make this "rocket science" accessible to everyone, explained in clear, simple language, with a touch of lightheartedness along the way. I promise there are no equations—although I may have planted one little gem in there. I wanted to break each section and chapter down into consumable chunks that stand alone, while fitting into a larger, richer tapestry spanning centuries, continents, science, philosophy, and more.

This is, at heart, the story of human potential, innovation, and our incredible ability to turn the impossible into reality.

Join me on this journey through space—not just as explorers of a new frontier, but as learners discovering better ways to live on Earth.

So now, to the stars…

SECTION 1: SCIENCE STUFF

Very Clever People

The night sky is vast and woven into our collective history. Across millennia, people turned constellations into calendars, tracked lunar cycles to govern harvests, and relied on celestial motion to steer caravans through deserts and ships across open water.

To ancient societies, the sky dictated the ebb and flow of life: when to plant, when to hunt, when to move, when to stay. From the Ziggurats of Mesopotamia to the stone circles of Scotland, architecture itself was aligned to the stars, out of reverence, and more importantly, utility.

What we now call *space exploration* began with that same motive: awe, romance, curiosity. The kind that builds tools, not just myths. The same curiosity that once tied a reed to mark a solstice now guides billion-dollar telescopes into orbit.

This section is our walkthrough of the history of how humans figured out how the skies work—and the scientific breakthroughs that made it possible to launch rockets, deploy satellites into orbit, and send things flying across our solar system.

Welcome to Science Class.

Chapter 1

Gods and Myths

Early Skywatchers

Before philosophers traced angles or priests carved equations into stone, the sky served as a practical calendar. In ancient Babylon, a priest ascended a stepped ziggurat to record the position of Venus against the night sky. With a sharpened reed, he pressed soft clay, fixing the planet's place against a backdrop of distant stars. Each mark tracked the timing of floods, the rise and fall of kings, and the rhythms of growing and harvest.

Patterns discovered on those tablets shaped the birth of astronomy. Babylonian scribes noticed that Mars would sometimes hesitate, drift backward, then lurch ahead—a riddle that kindled centuries of skywatching. Their base-60 system, crafted for trade and ritual, sliced the circle into neat fractions—sixty minutes, sixty seconds—that survive today in clocks and compasses.

Along the Nile, the ever-innovative Egyptians welcomed the first spark of Sirius at dawn, just before the life-giving flood of the river. Temple architects chiseled corridors to channel sunlight on the equinox, fixing their blueprints through the power of observation. Each stone became a measuring tool, set to mark the return of the star and the reliable swelling of the Nile's waters. The Great Pyramid of Giza, built in 2560 BC, was constructed with such precision that it points towards Orion, the hunter who was linked to Osiris, the god of the afterlife. Pretty amazing architecture to begin with—and incredible alignment.

While early starwatchers across the world tracked patterns in their own skies, in China, astronomers developed one of the world's longest continuous observational records. Court scholars documented comets, eclipses, planetary motions, and sudden "guest stars"–what we now call *novae* and *supernovae*–across more than four millennia. Their chronicles include the 1054 CE supernova that later formed the Crab Nebula, an event described with enough detail to allow modern reconstruction of its brightness and duration. Chinese observers also recorded sunspots centuries before the invention of the telescope, viewing the Sun through thin layers of smoked glass to protect their eyes. Their measurements of Halley's Comet enabled later astronomers to determine its orbital period, demonstrating how careful, long-term observation could reveal systematic patterns in celestial motion.

Thousands of miles away, Mayan astronomers watched Venus disappear and reappear, tracking its 584-day cycle with care and patience that spanned generations. Scholars used the position of Venus to schedule ceremonies and coordinate major civic events, folding complex calculations into bark-paper codices and preserving astronomical knowledge within their calendrical system. Their calendar systems stood unmatched until the Renaissance: the Maya tracked the orbit of Venus so precisely that they could forecast solar eclipses centuries in advance and charted irregularities in its path, even developing their own numerals for these astronomical records. Evidence of these achievements appears in their codices– history's earliest systematic program for predicting key celestial events and analyzing the stars.

Across the Pacific, Polynesian navigators set out beneath southern skies, reading the horizon for the rise of particular stars. They steered by night using the rising and setting points of stars, supported by knowledge of wind and wave patterns, and relying on memory to preserve routes between scattered islands.

In every civilization, skywatchers turned daily needs into innovation. They asked how, where, and when–marking the heavens with myth and measurements. These early observations became a scaffolding: a shared architecture for later astronomers who would build, question, and test against what the skies revealed each night.

This foundation of observation prepared the way for Greek scholars who sought underlying causes. Aristotle stood at the beginning of that effort, using evidence to shape the first comprehensive model of Earth and the heavens.

Aristotle and the Shape of the World

Aristotle (384-322 BCE) worked across many fields—logic, biology, physics, and the study of the heavens. He was a careful thinker who drew on observations gathered by earlier Greek astronomers and used them to build a structured picture of the world.

He paid close attention to clues that pointed to Earth's shape. During lunar eclipses, he watched Earth's shadow move across the Moon as a smooth, constant curve. On coastal hillsides, he noted how ships faded from sight on the horizon: the hull vanished first, then the mast, indicating the curvature of the sea's surface away from him. Travelers told him how the stars changed across latitudes, with familiar constellations slipping below the horizon as they moved south and new ones rising overhead. Taken together, these signs described a spherical world.

Aristotle placed that sphere at the center of the heavens. Around it he arranged a series of rotating spheres that carried the planets, the Sun, and the distant stars along steady circular paths. The system offered order, symmetry, and a sense of stability at a time when the sky's shifting patterns needed explanation. It became the framework many later thinkers inherited, influencing astronomy and philosophy for nearly two millennia.

Aristotle's influence reached far beyond his model. He showed how to build arguments from visible signs to draw conclusions about things that could not be reached directly. Modern astronomers still follow the same logic. They infer distant planets from tiny changes in a star's light or movement, reading faint signals from far away. Aristotle worked with limited tools, yet his approach established a way of learning from the sky that remains central to astronomy today.

Aristarchus: The First to Move the Earth

About a century after Aristotle, Aristarchus of Samos (310-230 BCE) proposed a model that shifted the entire frame of celestial thinking. He suggested that the Sun occupied the central position in the heavens and that Earth turned once each day while travelling around the Sun over the course of a year. The idea placed Earth in motion and removed it from the center of the system. People may have well thought he was mad.

Aristarchus reached his conclusions through geometric reasoning. During the half-moon phase, he measured the angle between the Sun and the Moon, using the geometry of right triangles to estimate their relative distances. He examined the shadow cast during lunar eclipses to compare their sizes. From these studies, he concluded that the Sun was far larger than Earth. A larger body, he argued, served naturally as the central reference point in the system, with smaller bodies circling it.

His estimates of sizes and distances fell short of our modern understanding, yet the method showed a disciplined approach: collect observations, apply geometry, and test ideas through proportion and scale. He worked within the tools of his time and used them to propose a structure of the heavens shaped by evidence rather than custom. It was, in essence, the beginning of a scientific approach to understanding the night sky.

Very little of Aristarchus's writing survives, and most of what is known comes from later references. Archimedes mentioned his heliocentric work with respect as one of several bold ideas. Ptolemy, whose geocentric scheme later dominated astronomical thought for centuries, did not address Aristarchus directly. As a result, the heliocentric model remained a lesser-known thread for many generations.

Eratosthenes: The Man Who Measured the World

Eratosthenes of Cyrene (c. 276-194 BCE), chief librarian at the Library of Alexandria, approached the study of Earth with a simple question: could the size of the planet be measured with the tools available to him?

He knew that in Syene—modern Aswan—vertical objects cast no shadow at noon on the summer solstice. Sunlight shone directly down wells, lighting their depths. At the same moment in Alexandria, a vertical stick produced a measurable shadow. Eratosthenes placed a gnomon—the pointer part of a sundial—upright and measured the angle between the stick and its shadow. The result was approximately 1/50 of a full circle, or about 7.2 degrees.

He treated Earth as a sphere, an idea supported by earlier Greek reasoning. If the Sun was far away and its rays reached Earth in parallel, then a difference in shadow angle between two cities must reflect a difference in their position along the curvature of the planet. The fraction of the circle represented by the shadow angle would match the fraction of Earth's circumference contained in the distance between Alexandria and Syene.

The next step was determining that distance. Eratosthenes used the best information available: reports in which professional bematists—trained distance walkers—had measured the separation between the two cities at roughly 5,000 stadia. The stadion varied by region, but modern scholars often assume the "Egyptian stadion" of around 157.5 meters, giving a total distance near 787.5 kilometers. Other interpretations use a stadion closer to 185 meters, producing a longer estimate, but the method remains the same.

With a shadow angle of 1/50 of a circle and a baseline of 5,000 stadia, Eratosthenes multiplied the distance by 50 and arrived at 250,000 stadia for Earth's full circumference. He later adjusted the value to 252,000 stadia, a number more convenient for division. Depending on which stadion length is used, his estimate lies remarkably close to the modern value of about 40,075 kilometers.

His approach relied on three components: a measured angle, a known distance, and the geometry of a sphere. From these, he produced one of the earliest successful calculations of Earth's scale. The accuracy of the result depended on assumptions about the stadion and the exact distance between the two cities, yet the method itself was rigorous and elegant. It demonstrated how simple observations, combined with reasoning and geometry, could reveal the size of the world.

Hipparchus: The Mathematician of the Stars

Hipparchus of Nicaea (c. 190-120 BCE) reshaped ancient astronomy. He approached the sky as a set of measurable patterns and treated earlier observations as data to be compared, tested, and refined. His work established methods that extended far beyond his lifetime.

He compiled one of the earliest detailed star catalogues, estimating the positions of hundreds of stars and organizing them by brightness. This produced the first known *apparent magnitude scale*, a simple system grouping stars into six classes from brightest to faintest. Modern astronomy still uses a refined version of this framework.

Hipparchus supported his cataloguing work with new mathematical tools. Drawing on geometric traditions and expanding trigonometric methods, he created early *chord tables*, which allowed astronomers to convert angular measurements into lengths and ratios. These tools improved predictions of celestial motion and supported more consistent geometric modelling.

His most influential insight came from comparing stellar records separated by more than a century. By examining older measurements from Timocharis and Aristyllus and comparing them with his own, Hipparchus detected a small but steady shift in the positions of the stars. The equinoxes—points defined by Earth's orientation as it moves around the Sun—appeared to drift westward along the ecliptic over time. The change was subtle, amounting to about one degree every seventy years, but it remained consistent across data sets.

Hipparchus recognized this as a real feature of the heavens and described it as the *precession of the equinoxes*. Modern astronomy interprets this phenomenon as *axial precession*: a slow, circular motion of Earth's rotation axis that completes a full cycle in roughly 26,000 years. The effect shifts the position of the celestial poles and changes which stars mark true north over long periods. Hipparchus did not explain the physical cause—Milutin Milanković (1879-1958), whom we will meet later, provided that explanation centuries afterward—yet his observational discovery remains one of the most important contributions in the history of astronomy.

He also improved eclipse predictions, refined estimates of lunar distance using parallax, and built geometric models that described the varying speed of the Sun along the ecliptic. These refinements later informed Ptolemy's more extensive system.

> **What is the ecliptic?** It's the invisible flat sheet traced out by Earth's orbit around the Sun. Seen from our viewpoint, the Sun appears to glide along this same great circle in the sky over the course of a year, and the other planets stay close to it because they orbit in nearly the same plane. The ecliptic also acts as a backdrop for the zodiac constellations, which anchored calendars, navigation, and myths for thousands of years. Tilted about twenty-three degrees relative to Earth's equator, it's the reason sunlight migrates north and south and we have seasons. And when the Sun, Earth, and Moon line up exactly in this ecliptic plane, we get solar and lunar eclipses.

Hipparchus established an approach that valued precise measurement, comparative analysis, and mathematical coherence. His star catalogue, magnitude scale, trigonometric work, and discovery of precession formed a framework that guided astronomical study for centuries.

Ptolemy's Circular Blueprint

Claudius Ptolemy (c. 100–170 CE), working in Alexandria, brought earlier Greek astronomy together into a single extensive system. His great work, later known as the *Almagest*, combined star catalogues, geometric models, and mathematical procedures into a reference that guided astronomical thought for more than a thousand years.

Ptolemy built on the measurements made by Hipparchus, such as tracking star brightness and recognizing the slow "wobble" of Earth's axis over time (precession). But he's most famous for creating a geometric model of the solar system that could predict where the Sun, Moon, and planets would appear in the sky.

In Ptolemy's model, every planet moved in a combination of circles. The largest circle—the *deferent*—traced out a broad path with Earth at the center. On top of this, each planet moved around a smaller circle called an *epicycle*. This arrangement let astronomers explain why certain planets, like Mars, sometimes seem to pause, move backwards (retrograde), and then resume their normal path across the sky—all while keeping Earth at the middle of the universe.

To get even better accuracy, Ptolemy added another component: the *equant*. Rather than having the center of a planet's circular path exactly on Earth, the equant was a separate point off to one side. This made each planet appear to speed up or slow down at just the right moments, even though the circles stayed perfect.

While the idea was a bit artificial, it worked: the model's predictions matched the sky well enough that Ptolemy's great book, the *Almagest,* became the go-to reference for navigators, astronomers, and calendar-makers for well over a thousand years.

Ptolemy also compiled a detailed star catalogue, assigning coordinates and magnitudes, and presented mathematical tables that allowed readers to calculate planetary positions for any given date. The clarity and completeness of the *Almagest* made it an enduring reference, and its predictive power helped sustain the geocentric model long after questions about its deeper truth began to emerge.

> **A Resting Place**: A rare Arabic edition of Ptolemy's *Almagest*, edited by the Persian astronomer Nasīr al-Dīn al-Tūsī in the 13th century, is preserved at the University of Edinburgh in Scotland. This edition is part of a significant collection of Oriental scientific works, a literary treasure entrusted to the university by John Baillie of Leys, whose family had long championed the preservation of astronomical knowledge. It is fitting that such an important work in the history of astronomy found a home in a place known for its rich scientific heritage.

While his system rested on assumptions that would later be overturned, Ptolemy's achievement was considerable. He gathered the observational and mathematical knowledge of his time into a structured whole, producing a model of the heavens that functioned effectively within the limits of available data. His influence shaped astronomical study throughout late antiquity and the medieval world.

Islamic Astronomers and the Foundations of Space Science

From the eighth to the fifteenth centuries, centers across the Islamic world transformed astronomy into a truly global enterprise. Great cities—Baghdad, Damascus, Cairo, Samarkand, Maragha—became hubs where scholars gathered, translating ancient Greek and Indian works, blending them with their own insights, and pushing mathematics and observation in new directions. What emerged were new tools, improved models, and a series of measurements that set the stage for centuries of discovery.

The House of Wisdom in Baghdad served as one of the earliest hubs of this activity. It functioned as a translation center, library, and research institution, where scholars collected and compared astronomical sources from multiple traditions. Among them, Muhammad ibn Musa al-Khwarizmi (c. 780–850) compiled detailed planetary tables and reshaped old astronomy with new mathematical models. His work formalized al-jabr, what we now call *algebra*, and introduced computational procedures that became the modern concept of the algorithm. These contributions provided astronomers with a clearer mathematical language for tracking and predicting the movements of the heavens. So, yes kids, algebra in school comes from ancient people in Baghdad figuring out space!

Further developments followed over the next centuries. Al-Battani (c. 858–929) spent decades measuring the sky—improving the accuracy of the solar year, refining values for the tilt of Earth's axis, and introducing updated sine and tangent tables that replaced outdated systems. His Zij al-Sab'i became the go-to reference for both Islamic and European astronomers for generations.

In the thirteenth century, Nasir al-Din al-Tusi (1201–1274) aimed to fix some of the stubborn problems embedded in the old Ptolemaic system. His most notable innovation was the *Tusi couple*, a geometric concept where a small circle rolls inside a larger one, converting steady circular motion into a straight-line back-and-forth motion. With the Tusi couple, astronomers could now model the planets' motions more accurately without relying on the awkward "equant" trick. This led to a theory that was more aligned with observations and simpler at its core. Under his leadership, the

Maragha observatory became a major center for mathematical and observational research.

A century later, Ibn al-Shatir (1304–1375) would push planetary modeling even further, replacing awkward equants with geometric tricks and creating models for the Sun, Moon, and planets that uncannily echo methods used by Copernicus two hundred years later.

During the fifteenth century, astronomical focus shifted eastward to Samarkand, where Ulugh Beg (1394–1449) built one of the world's most ambitious observatories of the pre-telescope era. Its central instrument—a massive stone sextant nearly forty meters long—enabled highly precise measurements of stellar positions. His star catalogue, the Zij-i Sultani, also known as the "Sultanic Tables," listed over a thousand stars and remained an authoritative source until the work of Tycho Brahe in the sixteenth century. Ulugh Beg also calculated the sidereal year with an accuracy remarkably close to modern values, differing by only a few seconds.

Over these centuries, Islamic scholars refined trigonometry, invented new instruments, updated star catalogs, and squeezed uncertainty from the cosmos bit by bit. Their ideas flowed westward, lighting the way for Copernicus, Tycho, and Kepler, and laying the mathematical foundation for the next great revolution in our understanding of the heavens.

Copernicus and the New Order

Nicolaus Copernicus (1473–1543) revisited a concept first proposed by Aristarchus more than 1,700 years earlier: the Sun, not Earth, serves as the central reference point of the planetary system. Working over many years, he assembled observations, geometric arguments, and mathematical analysis into a coherent heliocentric model.

Copernicus placed the Sun at the center of the system and set Earth in motion, rotating once each day and completing an annual orbit. This arrangement explained the apparent backward motion (retrograde motion) of planets such as Mars as a natural consequence of Earth's movement. It also simplified the order and spacing of the planets by relating them to their distances from the Sun.

He described his model in *De revolutionibus orbium coelestium* (*On the Revolutions of the Celestial Spheres*), published in 1543. The work retained several features of earlier schemes, including circular orbits and combinations of circles designed to match observed irregularities. It did not eliminate epicycles, yet it offered a new organizing principle that reduced the complexity of the system and placed planetary behavior within a uniform framework.

Reactions to Copernicus varied. Some mathematicians valued the clarity of the geometry. Others questioned the physical plausibility of a moving Earth, as daily rotation and annual motion could not be felt by people on the ground. The model entered scholarly discussion gradually and became one of several competing approaches to celestial mechanics.

The heliocentric system did not yet have the observational accuracy or physical explanation needed to replace the long-established geocentric tradition. Those refinements would come from later astronomers who built on Copernicus's structure. His contribution provided a new foundation: a model that related planetary motion to the Sun and opened a path for the deeper transformations that followed.

Chapter 2

And Yet It Moves

Maps, Orbits, and Telescopes

The transition from medieval skywatching to a mathematical understanding of orbital motion developed through improved instruments, systematic data collection, and the work of astronomers who challenged models that no longer aligned with observation.

Two figures in particular shaped this shift: Tycho Brahe, whose measurements provided the most accurate pre-telescopic data ever collected, and Johannes Kepler, who turned those measurements into the first predictive laws of planetary motion.

Tycho Brahe and the Discipline of Observation

Tycho Brahe (1546-1601) grew up within the Danish nobility but devoted his life to astronomy. A solar eclipse he saw in 1560 drew him toward the discipline, and in 1572, he recorded a bright "new star" in Cassiopeia—now known to be a *supernova*. Its sudden appearance contradicted the long-held Aristotelian belief that the heavens were unchanging. Tycho noted its absence of parallax, concluding that the object lay far beyond the Moon. This event alone indicated that celestial space was not the fixed, immutable realm that classical cosmology described.

Brahe is often associated with the prosthetic brass nose he wore following a duel conducted years earlier, an incident sometimes linked to an academic dispute, but his legacy came from his instruments. Brahe worked before the telescope, so he relied on precisely crafted quadrants, sextants, and armillary spheres—some several meters across—to measure angular positions as accurately as possible.

In 1576, funded by the Danish crown, he established Uraniborg on the island of Hven (now Ven, Sweden), a combined observatory, research center, and instrument-making workshop. Later, he built Stjerneborg, partially underground, to reduce vibration and wind disturbance. Over more than twenty years, Brahe measured planetary positions with unprecedented precision, achieving accuracies within one arcminute. Pretty impressive and something that would challenge and inspire his successor for years.

> **Quick Geometry Lesson**: a circle contains 360 degrees; each degree contains sixty arcminutes; each arcminute contains sixty arcseconds (or 1/3600th of a degree!)—for those paying attention, you will have noted the base-60 system came from the Babylonians! By example, the Moon spans roughly thirty arcminutes in the sky—half a degree. To measure planetary motion to within a single arcminute required instruments of extraordinary stability and an observer willing to repeat measurements across countless nights. Today, we use arcminutes to map the sky, map visual areas of the brain, and aim telescopes (and firearms), where it is called a "minute of angle" (MOA), with one MOA representing one inch of spread over one hundred yards.

Tycho's dataset was rigorous, but his cosmology still had a huge flaw. He proposed a hybrid model in which Earth remained stationary at the center, the Sun orbited Earth, and the other planets orbited the Sun. This Tychonic system kept many of the observational advantages of heliocentrism while avoiding theological objections to Earth's motion. But it did not resolve deeper structural issues. Rather, it offered a working idea at a moment when competing ideas were still being debated.

Kepler and the Geometry of Planetary Motion

Johannes Kepler (1571-1630) joined Tycho's team in Prague in 1600. He was mathematically skilled and inclined toward geometric models that sought order and proportion in nature. Tycho initially granted him limited access to the data, but after Tycho's sudden death in 1601, Kepler inherited his extensive planetary records, including a detailed series of observations of Mars.

Kepler sought to fit Tycho's measurements to circular orbits, continuing a tradition that stretched back to antiquity. But it just didn't work. Mars, in particular, refused to align with circular paths, even after Kepler applied epicycles and adjustments. Following the evidence, he replaced the circle with the ellipse, a shape defined by two foci rather than a single center.

Between 1609 and 1619, Kepler published three laws that described planetary motion more precisely than any earlier model:

Law of Ellipses (1609): Planetary orbits are ellipses, not perfect circles, with the Sun sitting at one focus. The closer together the two foci, the more the ellipse looks like a circle; pull them apart and the orbit stretches out. This explained why planets whip faster when they're close to the Sun and drift slower when they're farther away.

Law of Equal Areas (1609): Imagine a line connecting a planet and the Sun. As the planet moves along its orbit, this line traces out equal areas over equal amounts of time. Near the Sun, the planet speeds up; far away, it coasts. Kepler's second law is a great early example of the principle we now call conservation of angular momentum—even though no one knew the name back then.

Harmonic Law (1619): The square of a planet's orbital period (how long it takes to orbit once) is proportional to the cube of its distance from the Sun. This established a very important relationship between orbital size and orbital time, a key concept in modern celestial mechanics.

Kepler's laws removed the need for epicycles and concentric spheres. They offered a model built from geometric reasoning anchored in observational evidence. The laws later became essential to Sir Isaac Newton's formulation of universal gravitation, linking Kepler's empirical results to physical causes.

While Kepler was doing all of this paper-based homework, some other Europeans were about how to see the heavens for themselves.

The Telescope and the Opening of the Sky

While Kepler worked through Tycho's measurements, the telescope appeared in northern Europe. In 1608, lens maker Hans Lippershey applied for a patent on a device that "made distant objects appear near" by using convex objective lens and a concave eyepiece to magnify distant objects. Multiple craftsmen explored similar designs, but Lippershey produced functioning instruments and demonstrated their value to Dutch authorities, who recognized their military utility.

The new device spread quickly. Within a year, descriptions reached Italy, where Galileo Galilei constructed his own versions based on reports alone. By refining lens geometry, Galileo produced instruments capable of about twentyfold magnification—far beyond the power of the early Dutch devices.

Galileo the Astronomer and Physicist

Galileo Galilei (1564–1642) was a strong-minded Italian mathematician, physicist, and astronomer whose refusal to accept well-engrained beliefs radically changed the act of skywatching into a science. Early in his career, he taught mathematics and pursued studies of motion before turning systematically to astronomy.

One year after Hans Lippershey's invention of the telescope became known across Europe, Galileo built his own improved version. Using only second-hand descriptions and a bunch of experimentation, his best telescope could magnify objects twenty times—an enormous improvement over the original Dutch designs. While he didn't invent the telescope, he was the first person to systematically point it at the heavens and record what he saw. In 1609, he directed his improved telescope toward the sky and began recording observations that contradicted long-established assumptions.

In 1610, Galileo published *Sidereus Nuncius* (The Starry Messenger). It described a rugged lunar surface with mountains and valleys, four moons orbiting Jupiter, and the Milky Way containing innumerable stars. He also observed the phases of

Venus, which aligned naturally with a heliocentric model, in which Venus also orbits the Sun.

These findings challenged the Aristotelian view of perfect, unchanging heavens and provided real evidence that celestial bodies could move around centers other than Earth. Galileo's commitment to empirical evidence extended beyond astronomy. His experiments with falling bodies and inclined planes demonstrated that objects accelerate at the same rate in a vacuum regardless of mass, a principle foundational to later mechanics.

In a cathedral, Galileo noticed that chandeliers of the same length swung with consistent timing, no matter how wide their arcs. This concept, called *isochronism*, became the principle behind pendulum clocks, which revolutionized timekeeping.

Galileo supported the Copernican model openly. In 1616, the Catholic Church declared the Copernican model "formally heretical" and ordered Galileo to stop supporting it. He complied at first but sixteen years later in 1632, he published the *Dialogue Concerning the Two Chief World Systems* (1632). The book presented a debate between three characters discussing geocentrism and heliocentrism, with the pro-geocentric character named Simplicio. This thinly veiled mockery was not lost on the Church.

In 1633, Galileo was summoned before the Roman Inquisition, found "vehemently suspect of heresy," and forced to recant his views under intense pressure. He spent the remainder of his life under house arrest, forbidden to publish or teach. His writings were banned for centuries.

The phrase *"E pur si muove"*–"And yet it moves"–is often attributed to Galileo after his trial. There's no solid historical evidence that he actually said it, but the line endures because it captures the spirit of Galileo's defiance.

Galileo changed how people saw the stars and how they understood motion itself. He took the authority of understanding from religious texts and placed it in the hands of measurement and mathematics.

The four largest moons of Jupiter are known as the Galilean moons. A NASA mission that mapped the outer planets was named after him. Galileo set a new standard for how science is communicated—clear, direct, and aimed at the public. He was neither a saint nor a martyr, but a stubborn, brilliant storyteller driven by data. He refused to accept that Earth was still simply because it was said to be so. Galileo described the heavens and helped invent the very process we use to uncover the universe's secrets and understand why it moves as it does.

The Problem of Distance

Galileo's telescope cracked open the firmament, but even he couldn't measure just how far those stars really were. His instruments identified motion, yet distance remained a mystery.

Determining the distance to stars required a method that could detect extremely small shifts in their apparent positions. The principle behind it is the same effect you see if you hold your thumb at arm's length and alternately close each eye. Against a distant background, your thumb appears to shift sideways. This effect is *parallax*—an apparent change in the position of a nearby object when viewed from two different vantage points.

In astronomy, *stellar parallax* refers to the minute angular shift in a star's position when observed from two opposite points in Earth's orbit, typically six months apart. The size of that shift, measured in arcseconds, allows astronomers to calculate the star's distance using straightforward trigonometry. The method was known in antiquity, but the angles involved are so small—fractions of one arcsecond—that early instruments could not detect them.

Progress came in the nineteenth century. In 1832, Scottish astronomer Thomas Henderson (1798-1844) used precise instruments at the Cape of Good Hope to detect the parallax of Alpha Centauri. After returning to Scotland, he calculated a distance of about 3.25 light-years, close to the modern value of 4.37 light-years. Henderson delayed publication because he was uncertain about the residual errors in his measurements, and his result appeared after two others.

In 1838, Friedrich Wilhelm Bessel (1784–1846) published the first widely accepted stellar parallax, determining the distance to 61 Cygni. Shortly afterward, German-born Russian astronomer Friedrich Georg Wilhelm von Struve (1793–1864) announced a value for Vega. Henderson's Alpha Centauri result followed in early 1839. Although Henderson made the earliest successful observation, Bessel is credited with the first published and reliably confirmed measurement.

Stellar parallax became the basis for calibrating other distance-measuring techniques. Modern missions such as ESA's Hipparcos and Gaia satellites, along with the Hubble Space Telescope (HST), extend this method to much higher precision. Parallax also supports navigation for distant spacecraft: missions such as NASA's New Horizons use stellar parallax to estimate their position when conventional tracking becomes less effective at great distances.

A New Discipline Emerges

From Tycho's arcminute measurements to Kepler's geometrical laws and Galileo's telescopic observations, astronomy transitioned from a descriptive practice into a quantitative science. The motion of planets and stars became subjects of measurement, inference, and physical explanation. The groundwork laid by these astronomers set the stage for the next major step: a unifying framework that explained why planetary motion followed the patterns Kepler described.

That explanation would come from Sir Isaac Newton.

Chapter 3

Apples and Cannonballs

Sir Isaac Newton, the Main Man

By the late seventeenth century, the mysteries of the sky had unraveled much of their certainty. The old ideas of concentric celestial spheres had been discarded, circular orbits had given way to ellipses, and telescopes had shown us the heavens as rugged, uneven landscapes marked by craters and rough terrain. Yet one profound question remained unanswered: *why* did the planets move in such predictable ways?

Enter Sir Isaac Newton (1643-1727), a quietly brilliant wool-coated polymath—yet another one again—whose brilliance bridged everything from cannonballs to comets with one core idea: the same invisible force that makes apples fall to the ground is also what keeps the Moon in orbit around the Earth.

Born in 1643 in Lincolnshire, England—the same year Galileo passed away—Newton grew up immersed in the writings of great thinkers like Descartes, Galileo, and Kepler while studying at Trinity College, Cambridge.

In 1665, when the Great Plague forced Cambridge to close, Newton returned to his family home in Woolsthorpe for quarantine. Far from wasting the isolation, this period proved his *annus mirabilis*—his miraculous year—during which he laid the foundations for modern physics, optics, and mathematics.

The Birth (and Brawl) of Calculus

At that time, mathematics could describe lengths, areas, and volumes. But explaining how fast something was moving at any given instant–or how sharply a curve bends exactly at one point–was beyond its grasp. While most people stocked up on food during the plague, Newton created calculus. As you do.

Although he didn't call it calculus himself–that name came later from the Latin word for pebble–Newton described it as the study of "fluxions" (rates of change) and "fluents" (quantities flowing over time). He needed this new mathematics because algebra could not explain the curved paths of planets, and geometry alone couldn't capture the nuances of acceleration. Newton's calculus allowed him to slice motion into infinitely small pieces, each containing information about speed, force, and direction at that moment. This new approach changed motion from a set of snapshot points into a flowing, continuous process, giving science the language it needed to describe how things change over time. And a total headache for high school kids to come. And he wrote it in Latin!

Meanwhile, across the English Channel in Germany, the philosopher and mathematician Gottfried Wilhelm Leibniz (1646-1716) was independently developing his own version of calculus. His notation–with its elegant symbols like $\int$ for integrals and d/dx for derivatives–proved more user-friendly and is the notation still in use today. This sparked a decades-long academic feud between British supporters of Newton and Continental followers of Leibniz, filled with accusations of plagiarism and fierce debate. Historians now believe Newton developed his ideas first, around 1665, but Leibniz published his work earlier, in 1684. Their differing approaches complemented each other: Newton viewed motion geometrically, while Leibniz treated it like a language. The feud ended long after both men had died, with Newton preserving his legacy and Leibniz's notation becoming the standard. Call it a draw. Or as Newton might have muttered through clenched teeth, a fluent defeat.

Today, calculus is the engine that powers modern life. Sorry, kids, it is kind of hard to ignore. It governs any system that changes with time, space, or force. Calculus describes how fluids flow, how planets orbit, how bridges bend beneath weight and wind, and how electric currents or heat spread. It drives predictive models in climate science, medicine, and economics, forecasting everything from disease outbreaks to market ups and downs. Calculus lets us calculate the curves of race cars speeding around corners or the perfect rocket trajectories soaring through the atmosphere. Where algebra tells you where you are, calculus tells you *how* you are moving, *where* you'll be next, and *how fast* you'll get there. Newton created it to answer a fundamental question: *why do planets move as they do?* The answer lay in understanding rates of change.

Newton's Laws of Motion

We know Galileo charted how things move and Kepler sketched out the paths of the planets, but Newton wrote the laws governing the universe. In 1687, Newton published *Philosophiæ Naturalis Principia Mathematica*, known simply as *The Principia*. Written in dense Latin and with lots of equations—not exactly light reading—this work completely rewired physics, reshaped astronomy, and shifted the intellectual center of Europe. A true epic!

The first part of *The Principia* presents Newton's three laws of motion, which became the foundation of modern mechanics.

The First Law of Motion, the principle of inertia, states that an object remains at rest or continues moving in a straight line unless acted upon by an external force. Basically, the universe prefers to keep things as they are unless nudged.

The Second Law of Motion relates force, mass, and acceleration, showing that the heavier the object, the more force you need to change its motion—a simple cause-and-effect relationship.

The Third Law of Motion, famous as "for every action, there is an equal and opposite reaction," formalizes the idea of balance; if you push on something, it pushes back just as hard, explaining how rockets propel themselves by expelling gas downward.

What made these laws revolutionary was their universality. For the first time, the same principles explained why an apple falls and why planets orbit the Sun. No more dividing science into separate rules for Earth and the heavens—physics became unified.

Though Newton formalized the mathematical framework of these laws, many had earlier glimpsed parts of these ideas. Aristotle believed objects naturally come to rest unless constantly pushed, a view dominant for nearly 2,000 years. Iraqi mathematician and astronomer Ibn al-Haytham (also known as Alhazen) suggested moving objects carry an "impressed force" keeping them in motion, while Galileo showed that without friction, objects continue moving indefinitely. René Descartes (1596-1650) coined the term *inertia*, but it was Newton who tied these threads together definitively.

Gravity, the Force That Binds the Universe

In the second part of *The Principia*, Newton introduced his Law of Universal Gravitation, which explains how objects with mass pull on each other. Simply, every object with mass—whether it's a planet, a star, or an apple—exerts a force that attracts every other object with mass. The strength of this pull depends on two things: how heavy (or massive) the objects are, and how far apart they are. The heavier the objects, the stronger the pull between them. But the farther apart they are, that pull weakens really fast, following the "square of the distance." This means that if you double the distance between two objects, the force of attraction drops to just one-fourth as strong.

This law showed that gravity is a universal force acting everywhere, connecting everything in the cosmos. This equation linked Earth and the heavens: the Moon's orbit, ocean tides, falling apples—all obeyed the same invisible force. Gravity, once a vague concept or divine push, was now a universal, quantifiable law.

The famous story of Newton watching an apple fall is often oversimplified; he didn't get hit on the head. But he himself acknowledged that observing a falling apple made him wonder why it always fell straight down and how that force might extend to the Moon and beyond.

Falling Cannonballs

In *A Treatise of the System of the World*, which was published posthumously in 1728 based on a preliminary manuscript Newton wrote in 1685, Newton presented a thought experiment to illustrate how orbits work.

If we were to fire a cannon horizontally from the top of a very tall mountain, where the cannonball lands would depend on its speed. At low speeds, it follows a curved path and lands somewhere on Earth. As the speed increases, the cannonball travels farther before touching down. At exactly 7.8 km/s (28,440 km/h), it falls toward Earth at the same rate that the planet's surface curves away beneath it–resulting in a circular orbit. Speeds greater than this produce elliptical orbits, and at 11.2 km/s–the escape velocity–the cannonball overcomes Earth's gravity and heads off into space.

This cannonball analogy is commonly used when talking about orbits, orbital mechanics, and apparent weightlessness.

For the first time, nature could be predicted. Newton's laws enabled scientists to calculate comet orbits, cannonball trajectories, tidal patterns, and the exact constellations visible on any given night. Nature had become understandable–a universe governed by rules, not whim or mystery.

Wobbles and Tides

Newton's work also showed us how Earth, Moon, and Sun influence each other. Beyond the debate about whether the Earth or the Sun was at the center of our solar system lies another important idea: the barycenter.

We all think the Moon simply orbits the Earth. In fact, the Earth and Moon actually orbit each other, around a shared center of mass–the barycenter. Because the Earth has so much more mass, this barycenter is about 4,670 kilometers from the center of the Earth which is about three-quarters of the way to the surface from the center. One way to think about this is to balance a hammer on your finger: the shared mass is the whole hammer, but the balance point–the barycenter–is way closer to the heavier head. In the same way, Earth "wobbles" slightly as both bodies move around this point, creating subtle gravity variations that affect satellites and spacecraft trajectories.

This Earth-Moon interaction also produces tides through a gravity gradient–the Moon's pull varies across Earth's diameter. The side facing the Moon experiences stronger gravity, creating an ocean bulge. Surprisingly, a similar bulge forms on the far side because Earth as a whole is pulled more strongly than the water on the far side, which gets "left behind." This occurs because lunar gravity is weaker on the far side and because the Earth-Moon rotation around their barycenter creates an outward centrifugal effect felt equally across Earth.

I am sure you have been on a spinning playground ride (we used to call them roundabouts!): the farther from center you are, the stronger the outward force you feel. In the Earth-Moon system, every point feels this outward force. On the Moon-facing side, lunar gravity dominates; on the opposite side, the centrifugal force prevails. This creates two simultaneous tidal bulges—one toward the Moon and one directly opposite.

While the Moon primarily drives tides, the Sun also contributes with about half the strength of the Moon. When the Sun, the Moon, and Earth align (new/full moon), they create higher "spring tides." When at right angles (first/third quarter), their effects partially cancel, producing milder "neap tides."

Local geography also shapes tides. Continental shapes, ocean floor topography, and bay configurations create unique tidal patterns. Nowhere is this effect more spectacular than in Canada's Bay of Fundy, where tides can rise and fall by up to sixteen meters. The bay's funnel shape amplifies the incoming water, and the timing of the tide matches the bay's natural "sloshing" rhythm—like pushing a swing at exactly the right moment.

So Newton's discoveries about motion, inertia, and gravity explained how celestial bodies move as well as Earth's ocean patterns and spacecraft navigation principles—quite a remarkable achievement during a pandemic lockdown centuries ago with only quills and candlelight.

A New Spin on Motion

Newton's laws revealed how objects move in straight lines and how gravity binds bodies across space, yet another dimension of motion required further refinement: rotation. Objects do not simply travel; they spin, twist, precess, and wheel through the universe. From a child's spinning top to a tumbling asteroid, rotation also has some rules of its own.

Newton's First Law introduced inertia—the tendency of an object to maintain its state of motion unless influenced by an external force. Rotating bodies follow a similar principle. Once in spin, they keep spinning until an outside influence, called a torque, alters their motion. This idea became known as *angular momentum,* a quantity that measures how an object rotates, how its mass is distributed, and how vigorously it carries that rotation through space.

While Newton laid the foundation, understanding the full mathematical picture of rotation needed new tools. Throughout the eighteenth and nineteenth centuries, brilliant mathematicians like Leonhard Euler, Joseph-Louis Lagrange, and William Rowan Hamilton built upon Newton's ideas, creating a rich language of motion. Euler crafted equations that show how rigid bodies spin around their axes; Lagrange reimagined mechanics using energy and constraints; and Hamilton developed an elegantly simple way to describe motion—so elegant that it even became crucial in quantum mechanics. Their collective efforts turned the concept of rotation from just an idea into a precise, reliable science you can predict with.

Angular momentum also carries a remarkable feature—it remains conserved. When no external torque acts, the total amount of spin in a system stays constant, even if the shape or arrangement of that system changes.

You will have seen a figure skater on ice pull their arms closer to their body, making them spin faster and faster. This happens not because they have gained energy, but because their angular momentum stays the same while the distribution of their body mass changes.

This principle governs far more than icy performances. It provided the language for how objects turn, how they balance forces as they spin, and how they carry rotational memory through space. Planets develop their subtle seasonal wobbles from conserved angular momentum. Spacecraft adjust their orientation using spinning wheels that trade rotation between components. Engineers rely on angular-momentum calculations when designing turbines, gyroscopes, and satellites. This created the foundation for everything from modern satellite attitude control to the slow, majestic cycles of Earth's climate—cycles that Milutin Milanković would examine next with an extraordinary amount of patience.

Milanković and the Astronomical Clock

Why does the Earth's climate swing between ice ages and warmer periods? Earth's climate carries a rhythm—not the daily turning of the planet or the yearly swing of the seasons, but a far slower pulse measured across tens of thousands of years. Explaining that pulse became the life's work of Milutin Milanković (1879-1958), a Serbian mathematician and engineer who thought that climate depended on Earth's motion around the Sun rather than in volcanic ash or drifting continents.

Beginning in 1911, Milanković painstakingly set out to calculate how sunlight—*insolation*—varied across the planet over immense stretches of time. His tools were Kepler's ellipses, Newton's gravitational laws, and a boatload of patience. In 1920, he published his first major analysis; in 1941, he completed his masterwork, *Canon of Insolation and the Ice-Age Problem*, a thick volume that identified three slow, predictable changes in Earth's motion:

Eccentricity describes how stretched—or how nearly circular—Earth's orbit becomes over roughly 100,000 years. The orbit never turns dramatically oval, but even small changes in shape alter how much sunlight different regions receive over long timescales.

Obliquity is the tilt of Earth's axis. It currently stands at about 23.4°, but it oscillates between 22.1° and 24.5° over 41,000 years. A greater tilt exaggerates the seasons; a smaller tilt softens them.

Precession is the slow, top-like motion of Earth's axis tracing a broad circle in space. This cycle runs every 23,000 to 26,000 years and changes the timing of the seasons relative to Earth's position in its orbit. Just like our friend Hipparchus of Nicaea (c. 190-120 BCE) had suggested.

Each cycle appears small on its own, yet their combinations create profound climatic consequences. When northern summers grow slightly cooler for many thousands of years—because orbital geometry denies them stronger sunlight—winter ice lingers. Snow accumulates. Glaciers extend. When the cycles align the other way, summers intensify, and ice retreats. Through this celestial dynamic, Earth shifts between ice ages and warmer intervals.

Milanković's theory faced initial skepticism. Climate change on such epic scales seemed too complex to hinge on orbital geometry alone. Yet as twentieth-century science gained new tools—deep-sea sediment cores, isotope records, and detailed chronologies—the astronomical clock revealed itself. In 1976, George Hays, John Imbrie, and Nicholas Shackleton published a landmark study showing that the timing of ice ages over the past 450,000 years aligned with Milanković's predicted cycles. It was a dramatic vindication—not only could Earth's climate be traced in ocean sediments, but the mechanism itself matched the predictions of Newtonian and Keplerian physics and Milanković's calculations.

Today, these long rhythms are known as the Milankovitch Cycles, a cornerstone of paleoclimatology. They set the slow backdrop against which ice sheets breathe in and out over geological time. They unfold gradually, over tens to hundreds of thousands of years. Human-driven warming, by contrast, is unfolding over decades. The astronomical clock still turns, steady and indifferent, yet the pace of modern change is now far faster than anything in the cycles Milanković devoted his life to calculating.

The Unfolding Universe

By the turn of the twentieth century, Newton's universe looked complete. His laws traced the paths of planets, cannonballs, tides, and even the slow breathing of ice ages through the rhythms Milanković uncovered. Angular momentum explained how worlds spin, rotate, and hold their courses. Classical mechanics felt like a finished system.

But small cracks had begun to show. Mercury drifted in its orbit by an amount no equation could justify. Light behaved as though it played by different rules. Precision clocks hinted that time itself might not be universal. These anomalies were tiny but persistent—and they suggested that something deeper shaped the cosmos.

Solving that puzzle required someone willing to redraw the very stage on which the universe unfolds.

Chapter 4

Relativity

Einstein and the Shape of Space-Time

There was no way we were *not* going to talk about Albert Einstein (1879–1955), the famous wild-haired genius whose ideas bent space and time, and our entire picture of reality.

Einstein fundamentally changed the world of physics in a way few others have since Newton. His contributions didn't replace previous ideas but built upon them, leading to a new understanding where space and time are connected, matter and energy are interchangeable, and gravity is seen as a result of geometry rather than just a force. These ideas revolutionized astronomy, supported the development of modern cosmology, and provided important tools that continue to guide space exploration today.

Einstein grew up in Germany with a deep curiosity about light, clocks, and the structure of the universe. In 1905, his *annus mirabilis*, he produced four papers that altered scientific understanding at a stroke. Among them was the *Special Theory of Relativity*, which proposed that space and time form a single interconnected framework, and that the laws of physics remain the same for all observers moving at constant speeds. This theory introduced a world in which lengths contract, time dilates, and the speed of light remains constant regardless of the motion of the observer.

Special Theory of Relativity

Before Einstein, scientists assumed that time flowed at the same rate everywhere and that space had a fixed geometry independent of motion. Special relativity replaced those assumptions with a new structure in which measurements of length and time depend on an observer's speed. At everyday velocities, these effects remain imperceptible. Near the speed of light, they define the behavior of matter and radiation.

Einstein introduced the famous equation that has since become a shorthand for the modern universe: $E=mc^2$. (Sorry—an equation, but I thought this one at least was ok to include. You have heard of it, right?)

"E equals m-c-squared" means energy equals mass multiplied by the speed of light squared, and it tells us that mass and energy are two sides of the same coin—each can be converted into the other. This relation expresses the equivalence of mass and energy. A small amount of mass contains a vast amount of energy because it is multiplied by the square of the speed of light, an immense number. The principle explains why nuclear reactions release so much energy.

Mass-energy equivalence governs natural processes across the cosmos. Stars shine because hydrogen nuclei fuse into helium in their cores, releasing energy that travels outward as light and heat. In more extreme environments, such as the collapse of massive stars, the concentration of matter becomes so intense that it forms a black hole. Within the boundary known as the event horizon, ordinary descriptions of matter no longer apply; space and time curve inward with such intensity that no signal escapes.

Einstein's work on energy, mass, and motion created a language that engineers continue to use when designing nuclear power systems for spacecraft. Missions that travel far from the Sun, where solar panels produce too little electricity, rely on compact nuclear sources whose efficiency arises directly from the physics Einstein described.

The General Theory of Relativity

In 1915, Einstein extended his earlier work into the *General Theory of Relativity*, a new description of gravity. Einstein showed that mass and energy actually warp the fabric of space and time itself. The presence of mass curves space-time, like placing a heavy bowling ball on a trampoline, causing smaller objects to follow curved paths around it. A planet orbits a star because the straightest path in this warped geometry is an ellipse carved into space and time.

This theory produced predictions that soon found observational support. Starlight bends slightly as it passes near the Sun. Clocks run at different rates depending on their position in a gravitational field. The universe expands over time. Black holes arise as natural solutions to Einstein's equations.

Many of these predictions now play essential roles in space exploration. The Global Positioning System (GPS) relies on synchronized clocks in satellites high above Earth. Because time runs faster in weaker gravitational fields, GPS must apply corrections predicted by general relativity. Without them, positional accuracy would drift rapidly.

Einstein's equations also clarified how gravity influences spacecraft trajectories. Mission planners account for gravitational curvature when calculating orbital transfers, flybys, and deep-space routes. They also use relativity to evaluate how high-energy radiation behaves around massive bodies, informing the shielding required for astronauts and instruments.

Einstein once thought the universe ought to be static and added the *cosmological constant*—an extra term in his formulas meant to balance everything out. When later observations revealed that the cosmos expands, he reportedly called this addition his greatest misjudgment. Ironically, a modern version of the cosmological constant now helps describe the accelerated expansion of the universe.

General relativity offered a unified structure: matter shapes space-time; space-time shapes motion; and the interplay governs systems from planetary orbits to the large-scale evolution of the cosmos.

Relativity and the Path to Space

Together, Einstein's theories reframed the architecture of the universe. Space and time became dynamic, interdependent, and responsive to matter and energy. Nuclear reactions gained a clear theoretical basis. Gravity gained a geometric interpretation that explained planetary motion, stellar evolution, and the structure of galaxies.

These ideas extend directly into the modern era of spaceflight. Every satellite, deep-space probe, and interplanetary mission relies on systems governed by relativistic principles. From the timing of GPS signals to the curvature of trajectories, the imprint of relativity is everywhere. Einstein demonstrated that the universe operates through connections among space, time, mass, and energy. In doing so, he provided the conceptual framework for technologies that guide spacecraft, measure the cosmos, and support humanity's expanding reach beyond Earth.

So, with gravity and energy in hand, we face the next big question: how do we harness and control energy in the machines we launch into space? Let's explore some thermodynamics.

Chapter 5

Tea Kettles

✦ ·◆· ✦

Foundational Principles

To understand how rockets launch and reach space, we need to explore three key scientific areas: thermodynamics–the science of heat and energy; fluid mechanics–how liquids and gases move; and aerodynamics–the study of how air flows around objects. Together, with a bit of chemistry and physics, these fields form the foundation of rocket propulsion and spaceflight.

We are going to go back and meet some of the famous chaps who helped evolve the physics and science that led us to optimizing rockets. Don't worry–this will stay mostly conceptual. We'll just lay the groundwork before diving into how rocket engines actually work later.

The First Clues About Heat

Long before anyone wrote equations for rocket engines or calculated escape velocities, people were captivated by simpler mysteries: kettles whistling on stovetops, steam hurling iron across factories, pipes bulging with pressure, and locomotives shaking railway platforms to their bones. These were the first places where humanity watched invisible forces become visible: heat turning into motion, motion turning into pressure, and pressure turning into noise, chaos, and occasionally lawsuits.

Heat was once thought to be a kind of spirit that moved through the world like an unseen ghost, warming one thing and cooling another depending on its mood. You can imagine why: a pot heats up when placed over a flame but cools again in the air; metals glow when hammered; steam bursts out of kettles as if it wants to escape the tyranny of boiling water. Before physics, this was all very mystical.

Thermodynamics emerged from this daily experience of heat and motion and from engineers watching machinery rattle itself apart. Rockets still rely on exactly the same sequence of events as a boiling kettle: heat builds, pressure rises, gas searches for escape, and the whole system moves in the opposite direction.

This chapter is the story of how we learned to decode, predict, and master that process. It follows the earliest tinkerers exploring the wild nature of heat and the later engineers wrestling those same laws to hurl machines into space—each at the mercy of the rules that govern the simplest kitchen pot.

Understanding Gases

The first meaningful step toward rocket science began in the seventeenth century thanks to Robert Boyle (1627-1691). Experimenting with air pumps, Boyle noticed that if you squeeze air into a smaller space, it pushes back harder—just like when you try to compress a bicycle pump. In 1662, he wrote this down formally: pressure and volume vary in perfect opposition, as long as the temperature stays steady. This became known as Boyle's Law.

A century later in Paris, Jacques Charles noticed that balloons behave differently in winter: the cold shrinks the gas inside, and while warmer air revived them. Though Charles didn't publish his findings, French chemist and physicist Joseph Gay-Lussac (1778-1850) later confirmed and credited him.

And then in 1802, Gay-Lussac told us that at constant volume, gas pressure is directly proportional to temperature. A pressure cooker builds up pressure when heated—faster-moving molecules create more pressure in the same volume. This is often called Gay-Lussac's Law (though some refer to it as Amontons's Law, as Guillaume Amontons (1663-1705) made similar observations in the 1700s). This is why pressure cookers have safety valves to release excess pressure as temperature increases, and why pressure cookers whistle!

These early gas laws were simple enough, and they demonstrated a connection between pressure, temperature, speed, and the motion of unseen particles. However, something was missing.

In 1811, Amedeo Avogadro figured it out. He proposed that equal volumes of gas at the same temperature and pressure contain equal numbers of molecules—regardless of what those molecules are. His analogy would come much later but imagine two identical car parks: one full of small hatchbacks, the other filled entirely with large SUVs. Same capacity, same number of spaces. Avogadro's Number (6.022×10^{23}) represents molecules in one *mole* of substance, similar to how a car park might always contain exactly 100 spaces, regardless of what vehicles occupy them.

This was not obvious at the time. Many chemists still doubted that gases were made of individual particles at all. Avogadro was so far ahead of his era that his idea sat largely unnoticed for decades.

By the late nineteenth century, Wilhelm Ostwald supplied the term *mole*, formalizing the amount of substance required to turn Avogadro's idea into a working tool for chemistry and engineering. The mole is the amount of a substance containing as many entities (atoms, molecules, and so on) as there are atoms in 12 grams of carbon-12, establishing its role as a fundamental unit in chemistry.

The gas laws were now a family of insights. All that remained was to unify them.

The Ideal Gas Law

In 1834, Émile Clapeyron (1799-1864) took these separate threads—Boyle's pressure-volume rule, Charles's temperature-volume observation, Gay-Lussac's pressure behavior, Avogadro's particle count—and wove them into a single, compact equation. It related pressure, volume, temperature, and quantity of gas in a way that let engineers predict exactly how gases behave in pipes, cylinders, boilers, and, eventually, combustion chambers.

This was the Ideal Gas Law, a theoretical simplification, yes, but a really useful one. It told engineers that if you heat a gas, its particles race faster. If you confine that gas, pressure rises. If you let it expand through a narrow opening at high speed, it will push back on the container with equal force—creating thrust.

Every steam engine humming through nineteenth-century railways, every ship driven by boilers across the Atlantic, and every early internal-combustion engine owes something to Clapeyron. Rockets use the same thing, too.

When we speak of an engine "burning fuel," what we really mean is that chemical energy is turned into heat, heat becomes pressure, and pressure becomes motion. The Ideal Gas Law was the first mathematical language that let us predict this with confidence.

Kinetic Theory

The next breakthrough in the 1850s came when August Krönig (1822-1829) and Rudolf Clausius proposed that gases behave as they do because their tiny particles are always moving—rushing, bouncing, ricocheting off the walls of every container. Pressure in a gas wasn't some mysterious "force"—it's simply the collective slam of endless collisions. If the temperature rises, those molecules dart about even faster. If the temperature drops, they slow down. Now, gas laws snapped into focus: heat IS motion.

James Clerk Maxwell (1831-1879), one of Scotland's greatest gifts to science, took this further. He introduced the novel idea that gases are swarms of molecules moving at all different speeds, not the smooth, continuous fluids that earlier scientists had imagined. Picture a crowd in a busy train station. Most people amble along at an average pace. Some sprint for the train, and a few drift slowly, lost in thought often with headphones on. Maxwell showed us that temperature is simply the average kinetic energy of the molecular crowd. The feeling of "hot" is really just a reflection of how fast, on average, these countless particles are moving.

Austrian physicist Ludwig Boltzmann (1844-1906) then showed us how the universe's tendency toward disorder—entropy—is tied to the number of ways its particles can rearrange themselves. In other words, the messier the microscopic configuration, the higher the entropy. The more ways molecules can be shuffled, the greater the entropy. Disorder is simply more probable than order, so nature naturally drifts toward maximum entropy. This is important for when we get to the Second Law of Thermodynamics we will see shortly.

Maxwell also devised the Maxwell-Boltzmann Distribution, showing how molecular speeds spread out. Most cluster at moderate speeds, while a handful race ahead or lag behind. Temperature, in this new statistical view, was just a measure of random, individual motions—like the collective roar of a crowd made from countless shouts.

This statistical revolution did more than reframe our understanding of heat. It laid the foundation for rocket science: combustion means harnessing billions of frantic molecules, colliding, exchanging energy, and rushing out the nozzle at extraordinary speed.

The Three Laws That Rule the Universe

The First Law of Thermodynamics: Energy Never Disappears

Before rockets, engines, or even steam kettles, people argued endlessly about what heat actually was. In 1789, Antoine Lavoisier (1743-1794) proposed in his work, *Traité Élémentaire de Chimie*, that heat was a fluid called *caloric* that moved between objects, much like hot chocolate poured from one mug to another. It sounded good, but it was totally wrong.

Then came Sir Benjamin Thompson (1753-1814), better known as Count Rumford. In the 1770s, he suspected as much while watching cannon barrels being drilled produced heat in endless supply. If caloric were a real fluid, the supply should have run out long before the metal turned red hot. Rumford concluded that heat came from motion—the frantic jostling of particles—long before anyone could see atoms.

The English physicist James Prescott Joule (1818-1889), with his famous apparatus of falling weights turning paddles in water, demonstrated that mechanical work converts directly into heat. Work goes in, heat comes out, no mysterious caloric fluid needed.

Today, when you read about the energy stored in a chocolate bar, a rocket's fuel, or a stretched spring, you're really reading about joules. Scientists named the unit after Joule: one joule is the work done when a force of one newton moves an object one meter.

Not long after Joules, in 1850, Rudolf Clausius unified the concept as the First Law of Thermodynamics: energy can't be created or destroyed—only changed from one form to another.

In rocket science, this law is everything. A finite supply of chemical energy sits in each ounce of propellant. Ignite it—that energy becomes searing hot gas. Hot gas turns into pressure, pressure thrusts away as exhaust, and that exhaust is what launches us skyward.

Because tracking this energy gets messy, engineers use *enthalpy*, a measure of the total thermal energy of a substance—including the energy needed for it to exist at a given pressure. When hydrogen and oxygen combust, the change in enthalpy tells us how violently the gases expand and how much push we'll get out of the nozzle. A rocket engine is, at heart, an enthalpy management device with a very aggressive attitude.

The Second Law of Thermodynamics: Nature Always Takes a Cut

The nineteenth century was the age of steam. Engines were everywhere—and so was a great question: could they ever run perfectly, with no waste?

The French physicist and engineer, Sadi Carnot (1796-1832) answered firmly in 1824: no. Not even close. In his landmark treatise, *Reflections on the Motive Power of Fire*, he showed us that we could never turn all the heat from a burning fuel directly into work. There would always be some heat left behind.

Carnot imagined the perfect machine—an *ideal heat engine*, now known as the Carnot engine. This engine moves heat in a four-step process called the Carnot cycle: two steps where heat slides in and out at a steady temperature (isothermal), and two steps more where the engine simply coasts, neither gaining nor losing heat (adiabatic). It showed us that the efficiency your engine can ever hope to reach depends *only* on the temperature gap between where it grabs heat and where it dumps what's leftover–not on whether it burns steam, air, or the wildest chemical you can dream up.

Carnot also gave us the concept of *reversibility*, asking what would happen if you could, in theory, run the engine backward–a deep idea that became central to all of thermodynamics.

The Carnot cycle is still the benchmark for efficiency where no real machine, no matter how clever, can beat the Carnot engine running between the same two temperatures.

That brings us to Lord Kelvin (William Thomson) (1824-1907), the Belfast pioneer who wasn't satisfied with regular old thermometers. Lord Kelvin, who held a professorship of Natural Philosophy at the University of Glasgow for over 50 years, developed the absolute temperature scale (Kelvin Scale), which starts at absolute zero (-273.15°C, 0 Kelvin). This means that nothing can be colder than -273.15°C (0 Kelvin). He helped articulate thermodynamic laws showing that it is impossible to convert all heat into work and that heat engines have fundamental limits.

Rudolf Clausius followed in 1851, transforming physics with the Second Law of Thermodynamics: heat won't flow from cold to hot on its own, and the entropy of an isolated system–the measure of its disorder–always increases.

A decade later, Clausius coined the term *entropy* from Greek roots meaning "transformation" or "turning toward." He wanted a way to describe nature's tendency to move toward greater disorder. In his words, *"Die Entropie der Welt strebt einem Maximum zu"*–the entropy of the universe always heads for a maximum.

People often call entropy a "disorder," but it's more precise to say it measures the number of microscopic ways a system's parts can be arranged. The more possible arrangements, the higher the entropy. Nature drifts toward the states with the most possible arrangements, because statistically, that's what's most likely. In other words, nature tends to make a mess of things!

This is why coffee cools, engines waste energy, metals warp when heated, and why *no rocket* can ever be 100 percent efficient. Entropy is the universe's tax system. You can minimize the loss, but you can't eliminate it.

For rockets, the Second Law explains why no amount of engineering genius can stop waste heat from building up, why turbopumps need aggressive cooling, why propellant lines freeze, and why even the best engines lose energy in the churn.

The Third Law of Thermodynamics: Absolute Zero Is a Goal You Never Reach

By the early twentieth century, physicists wondered what happened as matter approached the coldest possible state. Walther Nernst (1864-1941) discovered that as you cool a perfect crystal toward absolute zero, its entropy—its internal disorder—approaches zero as well. Everything becomes so ordered that there is essentially only one microscopic arrangement left.

But Nernst also realized something else: reaching absolute zero is impossible. You can get close, but never all the way. Each step becomes harder than the last.

This became the Third Law of Thermodynamics, the final pillar of how energy behaves. Nernst's discovery set the ultimate temperature limit and explained why absolute zero is an unreachable goal. No matter how hard you try, you can get closer and closer, but you'll never actually arrive—much like chasing the speed of light.

Thermodynamics at the Launchpad

Every rocket launch begins with the same principles that make a kettle whistle: heat gathers, pressure rises, and gas seeks an exit. The only difference is scale. Replace a kitchen hob with a firestorm of chemical reaction, replace a spout with a finely shaped nozzle, and the same laws will lift a spacecraft beyond the atmosphere. From Boyle's pump to Maxwell's molecules, from Carnot's boilers to Nernst's frozen limits, we built a framework that lets us tame fire and cold in equal measure. Rockets exist because we learned to understand heat deeply enough to control it.

Rocket engineering is a daily confrontation with temperature. Cryogenic fuel tanks store liquids cold enough to turn steel brittle. Combustion chambers burn at more than 3,000°C–hot enough to melt the very engines that contain them, unless cooled correctly. Exhaust gases roar out at supersonic speeds, their pressure governed entirely by the principles Boyle, Charles, and Gay-Lussac uncovered centuries earlier.

The Ideal Gas Law predicts how the gases expand through the nozzle. Enthalpy tells engineers how much energy a fuel will release. The Second Law dictates how much of that energy can be turned into thrust. The Third Law sets the behavior of the fuel before it ever reaches the engine.

Liquid hydrogen must be kept cold enough to remain liquid, yet stable enough for pumps to drive it at enormous flow rates. Liquid oxygen, only slightly warmer, must be handled with equal care. Engineers design tanks with insulation that holds off the Sun's heat, valves that regulate pressure changes, and turbopumps that can handle a torrent of freezing fluids followed milliseconds later by combustion temperatures that rival stars.

The entire journey to orbit is shaped by thermodynamics: how efficiently fuel transforms into hot gas, how that gas accelerates through the nozzle, and how the engine manages the entropy it produces.

Rockets are thermodynamic machines. Their beauty comes from this interaction between heat and motion.

And now that we know how gases behave, we can turn to the next essential piece of rocket science: how fluids move when we force them through pipes, channels, pumps, and engines.

Chapter 6

Passing Gas

Rocket Plumbing

Long before a rocket ever roars to life, another world of motion has already begun. Fuel sloshes in tanks chilled nearly to absolute zero. Pumps spin hard enough to tear metal apart. Pipes carry cryogenic liquids, superheated gases, and everything in between. Turbulence forms, breaks, reforms, and occasionally attempts to ruin someone's afternoon.

If thermodynamics taught us why heat produces motion, fluid mechanics explains what that motion actually does. Every rocket is, at heart, an elaborate plumbing system that takes gases and liquids, persuades them to behave for a few minutes, and then releases unimaginable violence in a precisely controlled direction.

Fluid mechanics sits beneath every stage of modern spaceflight. It governs how fuel reaches the combustion chamber, how exhaust expands through a nozzle, how life-support systems circulate breathable air, and how coolant is routed through engines that would otherwise melt. It even tells us why astronauts must stay away from trapped pockets of carbon dioxide that stubbornly refuse to drift away in microgravity. Without it, nothing moves—not the fuel, not the exhaust, not the spacecraft.

This chapter explores how we learned to predict, tame, and exploit the way fluids behave. And it begins with the oldest of technologies: water flowing downhill.

From Rivers to Rockets

Ancient engineers learned fluid mechanics the slow and hard way—by watching nature misbehave.

The Egyptians didn't need equations to know what happens in the Nile's annual floods: they just built canals that went with the river's moods. Roman aqueduct builders knew that water could make a mess with pressure surges, bursting pipes and collapsing stonework. All across China, India, and Mesopotamia, whole civilizations flourished or failed based on their ability to channel water—predicting its path through fields, canals, and over cunningly placed spillways and sluices.

Though crude compared with modern physics, these early feats revealed that fluids don't care what you want. You can coax, redirect, or outsmart them with geometry, but once water (or fuel, or hot gas) gets moving, it becomes incredibly hard to really control.

Fluid mechanics has become the backbone of rocket engineering, where the stakes for getting it wrong include turbopump explosions, engine shutdowns, and occasionally the kind of "rapid unplanned disassembly" (aka *boom* time!) that makes headlines.

Dropping Pressure Quickly

In the early 1730s, Swiss mathematician and physicist Daniel Bernoulli (1700-1782) began experimenting to understand how fluids like water and air move. Inspired by his father Johann's work, he studied how water flowed through pipes of different shapes at the University of Basel. Bernoulli noticed that when water traveled faster through a narrow part of a pipe, the pressure pushing against the pipe's sides dropped.

I am sure this must have been a bit of an eye-opener. Most natural behavior suggests the opposite: push something harder, and the pressure rises. But Bernoulli saw that the total energy of a moving fluid stayed constant. If a fluid gained speed, that energy had to come from somewhere, and it came from its pressure.

After years of experiments and calculations, Bernoulli published his findings in 1738 in his book *Hydrodynamica*, introducing what we now call Bernoulli's Principle–the observation that when a fluid speeds up, its pressure drops.

This deceptively simple principle later explained how airplane wings create lift, how rocket nozzles work, and why perfume atomizers spray their mist.

For airplane wings, the wing's shape makes air travel faster over the curved top surface and slower under the flat bottom. According to Bernoulli's principle, the faster-moving air above has lower pressure, while the slower air below has higher pressure. This difference pushes the wing upward, creating lift.

For rockets, when hot exhaust gases zoom through a narrow nozzle, they speed up and their pressure drops. This change turns the energy in the exhaust into a strong forward push that propels the rocket. It's just like squeezing a garden hose–covering part of the opening makes the water speed up and spray out farther.

Atomizers, perfume sprayers, and even paint sprayers rely on Bernoulli's principle as well: when fluid velocity increases in a constricted path, pressure decreases, which draws liquid into the airflow and disperses it as droplets.

He also helped us understand the Venturi effect–the phenomenon in which a fluid forced through a constriction speeds up dramatically, dropping its pressure even further.

To understand *why* fluids accelerate the way they do, *how* forces distribute themselves inside a moving stream, and *what* governs the shape of every jet, plume, shock, and swirl, the next step came from one of the great architects of mathematical physics: Leonhard Euler.

The Perfect Fluid

Building on Bernoulli's work, Swiss polymath Leonhard Euler (1707-1783) made a major breakthrough in 1757. He showed how "perfect" (or "ideal") fluids (those without friction) would behave. While working at the Berlin Academy of Sciences, Euler developed mathematical tools that could predict fluid movement with amazing accuracy. Rather than treating fluids as collections of tiny particles, his approach viewed them as continuous flows and gave us the concept of *flow fields* that would later become fundamental when designing airplanes, rockets, and engines.

Of course, perfectly frictionless fluids don't actually exist in nature. However, Euler's equations gave engineers their first real mathematical tools for designing rocket components. His approach allowed scientists to analyze fluid flow at any specific point in space and time—a concept that became important for modern computer simulations of fluid dynamics. These equations work especially well for modeling how rocket exhaust gases expand in the vacuum of space, where the absence of atmospheric pressure makes the flow behave more like Euler's idealized model.

Euler's mathematics remains so central to spaceflight that if you removed it from engineering textbooks, nearly every chapter on propulsion, aerodynamics, and nozzle design would simply vanish.

But real fluids are not perfect. They stick. They swirl. They misbehave. And sometimes, they just make a mess!

Navier-Stokes' Sticky Business

The real breakthrough for practical applications came in the 1800s with the work of two brilliant scientists. Claude-Louis Navier (1785-1836), a French engineer and physicist, and George Gabriel Stokes (1819-1903), a British mathematician and physicist, tackled something previous scientists had ignored—the messiness of real fluids.

In 1822, Navier published his initial work on the equations of motion for "real" viscous fluids—those with a bit stickiness, like oil or honey.

In 1845, Stokes expanded and formalized Navier's equations, giving them more rigorous mathematical treatment and clarifying the physical meaning of viscosity. Together, their famous Navier-Stokes equations account for viscosity, which is simply the internal friction within a fluid. Think about how honey moves more slowly than water or how syrup sticks to a spoon—that's viscosity.

To understand how these principles work together, consider an aircraft wing. Bernoulli's Principle explains that faster-moving air creates lower pressure, while the Navier-Stokes Equations show how air actually flows and sticks to surfaces. As air moves over a wing's curved top, it speeds up (Bernoulli), while also forming a thin layer of slow-moving air right against the wing (Navier-Stokes). This creates a pressure difference between the top and bottom surfaces, generating and maintaining lift while keeping the airflow "attached" to the wing surface. This is actually how we minimize drag.

Spilling Your Wine

The next time you pour a bottle of wine and the liquid runs elegantly down the side and drips on the table, or a nice white tablecloth, instead of into the glass, you're watching Navier-Stokes at work. This is also called the *teapot effect*.

As the wine reaches the lip of the bottle, gravity tries to pull it outward in a smooth stream. But surface tension and viscosity have other ideas. The molecules prefer to cling to the glass, forming a thin layer that wraps around the curve and dribbles down the outside. The same sticky behavior—that balance between inertia and internal friction—is what Navier and Stokes captured in their equations for "real" fluids.

In miniature, the bottle lip behaves like an aircraft wing or a rocket nozzle: a thin boundary layer of fluid sticks to the surface while the faster flow above it moves cleanly away. The only difference is that, in one case, you get lift; in the other, you get stains.

These equations are incredibly powerful but also extraordinarily complex. In fact, fully solving them in all situations remains so challenging that it's one of the Clay Mathematics Institute's Millennium Prize Problems, with a million-dollar reward. Despite this mathematical complexity, they're essential tools for modern rocket design, helping engineers predict everything from how fuel flows through pipes to how plasma forms around spacecraft during reentry.

Reynolds Number

Later in the nineteenth century, Belfast-born engineer and physicist Osborne Reynolds (1842–1912), a professor at the University of Manchester, made a key discovery that would revolutionize our understanding of fluid dynamics. In 1883, Reynolds designed an experiment using glass tubes through which he pumped water. By injecting a thin stream of dyed water into the clear flow, he noticed that at low flow rates, the dye stream remained distinct and steady (laminar flow), but as he increased the flow rate, it would suddenly break up into swirls and eddies (turbulent flow).

Reynolds also noticed that this transition from smooth to chaotic flow wasn't random at all but rather it occurred at a predictable point that could be calculated using what we now call the *Reynolds Number*.

It is a way of understanding flow and chaos. An example might be ice skating on a quiet pond with just a few skaters as everyone glides smoothly around without bumping, following smooth paths. This is a low Reynolds Number. But now, you are in a packed ice rink in New York's Central Park, where people are cutting across each other, and the motion becomes unpredictable and chaotic. This is a high Reynolds Number and likely comes along with a few well-chosen superlatives!

Reynolds Number is calculated by comparing two forces: how fast the fluid (or skaters) wants to move forward (inertial forces) versus how much it wants to resist that movement due to internal friction (viscous forces). It is just a number that has become a fundamental tool for predicting when fluid flow transitions from laminar (smooth) to turbulent (chaotic).

Reynolds Number helps engineers design spacecraft systems— from fuel injection systems that need controlled turbulence for proper mixing, to heat shields that rely on smooth airflow for safe reentry. Even the air circulation system on the International Space Station (ISS) is designed using Reynolds' principles to ensure astronauts have properly mixed, breathable air without disturbing pockets of CO_2.

When Fluids Fight Back

Rocket engineers struggle with three troublemakers every day that haunt their design meetings: jetting, cavitation, and cryogenic flow instability.

Jetting starts when fluid races through a tiny opening and forms a high-speed jet—powerful enough to drill holes, carve metal, and throw a rocket off balance. It is like piercing a can of soda or beer after you have shaken it and watching the liquid rocket out of the tiny hole in a sharp stream—imagine the force behind that if scaled up to rocket engines. In rocket plumbing, injectors use tiny nozzles to fling fuel and oxidizer into the combustion chamber. If the jetting gets too wild, streams blast around like punctured soda cans, shaking the engine and setting up pressure waves that rattle the entire rocket. Tone it down too much, and you're left with a sluggish flow leading to poor mixing, weak thrust, and dangerous hotspots.

Cavitation is even sneakier. Think of dropping a hot pan into cold water and hearing that furious hiss as vapor bubbles form and collapse. When pressure drops fast enough inside rocket propellant lines, liquid boils into clouds of tiny vapor bubbles. As these bubbles implode, they hammer metal like an army of microscopic mallets, tearing into turbopump blades and destabilizing the flow. The mighty Saturn V F-1 engine wrestled with cavitation during its development, and today's engineers still lose sleep over it. Even when engineers tame cavitation's microscopic hammer blows, cryogenic propellants introduce an entirely different breed of mischief, given the extreme temperature involved.

Cryogenic instability is the final wildcard. Imagine a full, unopened can of soda—cold, sealed tight, and under pressure. As long as it remains undisturbed, everything stays quiet. But the moment you expose it to heat, shake it, or open the tab, unpredictable things happen: bubbles form, pressure builds, and the contents can fizz or spray out unexpectedly. You never know if it's going to quietly hiss or erupt into a wild, messy explosion.

Cryogenic fuels like liquid hydrogen behave much the same way, but on a far more challenging scale. Stored at ultra-low temperatures and under pressure, they're remarkably stable—until the tiniest temperature change, vibration, or disturbance causes sudden bubbling, sloshing, or rapid boil-off. Like a shaken can, these cryogenic fluids can go from calm to chaos in an instant, making them tricky to control in rocket engines. Rocket designers must manage all this behind tough metal walls, because just like with a soda can, what happens inside often stays hidden—until it doesn't. Early SpaceX Starship tests highlighted this challenge vividly, as methane tanks experienced pressure spikes and "geysering" during rapid chill-in, where warming pockets of cryogenic liquid expanded unpredictably inside feed lines. We will meet geysering again later.

These three bad boys are the reason rocket plumbing is among the most challenging engineering fields on Earth.

The Digital Wind Tunnel for the Space Age

Modern rocket design relies on computer simulations called Computational Fluid Dynamics (CFD) to understand how liquids and gases flow through rocket systems before anything is physically built. Engineers have the incredible ability to simulate a wide range of phenomena. They can model everything from how fuel mixes inside engines, to how air circulates in crew cabins. They also predict how shockwaves might influence a spacecraft during reentry, and work to optimize life-support airflow in the ISS—taking into account the unique environment without gravity, where exhaled carbon dioxide can collect in pockets around astronauts' heads.

They also save enormous amounts of time and money by catching potential problems early. Engineers can watch how fuel will burn, predict heating patterns during reentry, and visualize airflow around the vehicle—all on a computer screen.

Space has some really unique challenges since fluids behave differently without gravity. Fuel doesn't settle at the bottom of tanks, and air doesn't naturally rise or fall. Engineers apply these same scientific principles but adapt them for zero-gravity, creating specialized systems for fuel management and air circulation that work in the weightless environment.

Who knew passing gas could be so well understood?!

Chapter 7

Shocking Turbulence

Ripping Through the Sky

By the time fuel has raced through pipes and turbopumps and exploded out the back of a rocket in a superheated fury, the vehicle faces a far simpler adversary: air. Not the gentle breeze you feel on a hilltop in the highlands, but an atmosphere that behaves like a living thing—thick, thin, violent, predictable, unpredictable, and not happy with the speeds you are attempting. A rocket climbing toward orbit has to deal with this atmosphere every meter of the ascent. It bends it, shoves it, slices it, compresses it, and is punished for every miscalculation along the way.

Fluid mechanics explained how gases behave *inside* the rocket. Aerodynamics explains how gases behave *around* it. And if thermodynamics is the story of heat, aerodynamics is the story of motion through chaos.

Humanity did not begin by imagining rockets streaking through the sky at Mach 20 (twenty times the speed of sound). We began with windmills creaking in village fields, sails filling with the breeze, and wooden gliders skimming along dunes in North Carolina.

This chapter follows that journey—from the first hints of airflow to the brutal physics of rockets ripping through the sky.

The Four Forces of Flight

Every object moving through the air—whether an arrow, a hang glider, or a 3,000-tonne rocket—lives under the rule of four forces.

- **Weight** is gravity's steady grip that pulls everything downward.
- **Lift** is the upward force caused by the flow of air around a surface that creates pressure differences. This is what wings do!
- **Drag** is that force that counters forward movement, caused by a negative pressure off the back of a moving vehicle.
- **Thrust** is the force that pushes the object forward—like a swimmer kicking off a pool wall, an aircraft driven by its engines, or a rocket hurled upward by fire exiting out its backend.

An aircraft, whether they're banking into a turn or cruising straight and level, constantly balances all four forces: weight, lift, drag, and thrust for stable flight. However, rockets are different.

During launch, their engines must produce a massive amount of thrust to outmuscle their weight (downward force caused by their sheer mass and gravity) and punch through the air resistance that increases as it accelerates. Their long, narrow shape helps minimize drag, and their side fins produce some stability—just not in the same way wings do for an aircraft.

As the rocket streaks upward, air gets thin and drag drops away. Fins lose their bite in the thinner air, and only thrust really matters. It's only above the thickest atmosphere that the engines can finally open up, racing toward the speeds that separate orbiters from aircraft. Gravity is always there, but for true spaceflight, the real wall is speed. To stay aloft in orbit requires velocity not just altitude—blazing around Earth at about 28,000 kilometers per hour (7.8 km/s)—a feat only achieved after a rocket has survived everything the lower atmosphere can throw at it.

Max-Q

No matter the size, shape, nationality, or mission, every rocket faces a single moment that defines its launch: Maximum Dynamic Pressure, or Max-Q,.

Max-Q strikes about sixty to ninety seconds after liftoff, when the vehicle has picked up speed—fast enough for the thick lower atmosphere to throw its strongest punch. The rocket is hurtling skyward, not yet high enough to escape dense air, but racing so quickly that aerodynamic forces slam it from all sides. Imagine sticking your hand out the window of a car travelling at motorway speeds—now imagine the same at 1,000 km/h with the rocket carving through thick air. That is Max-Q: the moment the rocket must plough through its greatest aerodynamic punishment.

At this point, the rocket faces its greatest structural challenge. The dense air together with the rocket's velocity builds up the largest aerodynamic stresses anywhere on the flight. Engineers plan for this. When sensors report Max-Q approaching, engines throttle down—briefly holding back their power to make sure the rocket's skin and bones aren't ripped apart. When Max-Q has passed and the air thins, the engines surge again, and where it is less inclined to be snapped in half.

But surviving Max-Q isn't just about brute strength. It requires understanding something invisible, fragile, and utterly essential: the boundary layer.

The Boundary Layer

In 1904, Ludwig Prandtl—another brilliant polymath—revealed the hidden structure of airflow. When air moves over a surface, not all of it slips by at the same speed. A thin layer right next to the surface slows down dramatically due to friction. This fragile film, only millimeters thick, is called the boundary layer, and it determines everything that happens in high-speed flight.

If the boundary layer stays smooth and orderly (laminar), drag is modest and predictable. If it becomes chaotic (turbulent), drag spikes and the airflow becomes very messy. A rocket would love nothing more than a laminar boundary layer from launch to orbit. The atmosphere has other ideas.

As rockets approach transonic speeds—near Mach 1—the boundary layer becomes unstable, buffeted by shock waves that form and collapse around the vehicle. Parts of the airflow turn supersonic while others lag behind, causing shock waves to appear up and down the rocket, rattling the vehicle and threatening to rip the boundary layer away entirely.

At supersonic and hypersonic speeds, the atmosphere becomes more aggressive still. During re-entry, shock waves compress the air so severely that temperatures skyrocket—hundreds or even thousands of degrees—turning the air into a glowing plasma sheath. We will hear more about this a few times over a number of chapters.

The boundary layer really is the rocket's shield but can also be a massive problem child. It must be managed through a combination of material science, shape, speed, structures, and understanding where you are in the atmosphere.

Introducing Mr. Hypersonic

In the early 1950s, Theodore von Kármán (1881–1963), a Hungarian polymath—these darn clever polymaths!—who later worked in the United States, was already well-known for his groundbreaking work in aerodynamics and rocket propulsion at the Jet Propulsion Laboratory (JPL)—and was a founding figure at JPL and Caltech. His research had focused on supersonic and hypersonic flight, helping develop the mathematical foundations for high-speed aircraft design. While studying the behavior of vehicles at extreme altitudes, he asked a fundamental question: where does the atmosphere really end? And space start?

von Kármán discovered that at around 270,000 feet, the atmosphere becomes so thin that conventional aerodynamic flight ceases to be viable. At this altitude, a winged aircraft would need to reach nearly orbital speed to stay airborne—that's more like orbiting Earth than flying through air. As atmospheric models improved, he later increased his estimate slightly to just over 280,000 feet, about 80 to 85 kilometers.

The now well-known 100-kilometer mark, now known as the Kármán Line, was later adopted by international record-keeping organizations as a clear, standardized boundary, and officially recognized by the Fédération Aéronautique Internationale (FAI) in the 1960s. Meanwhile, the United States awarded astronaut wings at 50 miles, which is closer to von Kármán's original physics-based estimate.

This invisible line marks the transition from aerodynamics to astronautics. Below it, lies the regime of lift, drag, shock waves, and atmospheric flow; above it, these forces lessen, and orbital mechanics take over. Every rocket has to pass through the first regime before fully entering the second.

Aerodynamics Across the Solar System

These aerodynamic principles apply across our solar system, each environment presenting unique aerodynamic challenges. On Mars, where the atmosphere is about one percent as dense as Earth's, engineers had to completely rethink parachute design.

The Mars Perseverance rover needed a special parachute to land safely. This parachute was huge—about seventy feet across (21.5 meters)—making it one of the biggest ever sent to Mars. Engineers built it using strong materials like nylon and Kevlar.

Unlike parachutes on Earth, this one had to open while traveling incredibly fast—about 1,300 to 1,500 kilometers per hour (about Mach 1.7). That's much faster than a commercial airplane! This extreme speed was necessary because Mars's atmosphere is so thin that the parachute needed to catch enough of this thin air to slow the rover down before landing.

If Mars barely has enough air, Venus has far too much. Its atmosphere is ninety-two times denser than Earth's at sea level and hot enough to melt lead. The Soviet Venera probes that visited Venus were built like deep-sea submarines, with strong hulls to handle the crushing pressure.

Future Venus missions like NASA's DAVINCI+ and Europe's EnVision will use specially designed parachutes, heat shields, and aerodynamic shapes to help spacecraft descend safely. While turbulence does exist in Venus's atmosphere, the biggest challenges are actually the extreme pressure, scorching heat, and highly corrosive atmosphere. Engineers focus on creating descent systems that can handle these harsh conditions while maintaining a controlled, slow journey to the surface.

Even in the apparent vacuum of low Earth orbit, aerodynamics plays a really important role. The International Space Station (ISS) orbiting at 400 km up, encounters enough residual atmosphere to create drag that slows it down by about two kilometers per month. To counteract this, the ISS uses regular reboosts from visiting spacecraft or its own thrusters. These burns typically increase the station's velocity by a few meters per second, just enough to maintain its orbital altitude. Without these corrections, the ISS would eventually spiral back into the atmosphere, much like a satellite at the end of its life.

This thin upper atmosphere, known as the *thermosphere*, also affects satellite operations. Solar activity can heat and expand this layer, increasing drag on satellites and potentially changing their orbits. Space agencies and satellite operators must constantly monitor and adjust for these effects to maintain exact orbital positions. Space can be such a drag!

Every modern spacecraft, from SpaceX's Dragon to NASA's Orion, relies on these aerodynamic and fluid mechanic principles. They guide everything from launch trajectories to heat shield design. While we may dream of the vacuum of space, we must first master the art of moving through air—our first and last challenge in every space mission.

The Violent Business of Reentry

Reentry is the ultimate test in high-speed aerodynamics. A returning spacecraft is essentially a controlled meteor, slamming into the atmosphere at speeds that can exceed Mach 25. The compression of air ahead of the vehicle heats it to plasma temperatures. Spacecraft glow. Heat shields char. Shock waves wrap around the vehicle like a second skin. We will talk more about this later.

All of this high-speed aerodynamics—the shape of the shock, the thickness of the boundary layer, the distribution of heating—is governed by the same principles formulated by Bernoulli's pressure drops, Euler's perfect fluids, Navier-Stokes viscosity, Prandtl's boundary layers, and von Kármán's hypersonic logic.

The physics that guide your hand out a car window become, at these speeds, the physics that determine whether astronauts come home safely. They tell engineers how to shape a rocket's nose, how to cool a heat shield, how to predict shock waves, how to steer at Mach 5, and how to survive the screaming chaos of reentry.

And now, with thermodynamics, fluid mechanics, and aerodynamics firmly in place, we can move onto understanding some of the science behind rocket fuels and getting into orbit.

Chapter 8

Rocket Fuels

Burning Stuff Up

Rocket propulsion has come a long way since its earliest days. Different types of rocket fuels each have their own strengths that make them useful for specific missions. Let's briefly explore the major categories of rocket fuels, how they developed over time, and what makes each one special.

No history of rocket fuels would be complete without mentioning John Drury Clark (1907–1988), an American chemist and rocket propellant expert whose work and writings dramatically influenced rocket propellant development. His book *Ignition! An Informal History of Liquid Rocket Propellants* (1972) remains a classic in the field.

Clark worked during the post-WWII era when rocket scientists were frantically searching for higher-performing propellants. He tested many exotic and often extremely dangerous substances, earning him legendary status among propulsion engineers.

Chlorine trifluoride was just one of many hazardous substances Clark worked with in his quest for better rocket fuels. In Clark's own memorable words: "It is, of course, extremely toxic, but that's the least of the problem. It is hypergolic with every known fuel, and so rapidly hypergolic that no ignition delay has ever been measured. It is also hypergolic with such things as cloth, wood, and test engineers, not to mention asbestos, sand, and water—with which it reacts explosively."

The Reliable Workhorse

The classic combination of kerosene (RP-1) and liquid oxygen has powered everything from early missiles to modern commercial rockets. Refined in the 1950s, RP-1 is a highly purified form of kerosene specially designed for rockets. When kerosene meets liquid oxygen, they react vigorously to produce carbon dioxide, water vapor, and tremendous energy.

One big advantage is that kerosene can be stored at normal temperatures, unlike fuels that need extreme cooling. It's also quite dense, allowing for smaller fuel tanks, which means rockets can be more compact and efficient. This combination provides good thrust with moderate efficiency, making it ideal for first-stage rockets.

The Saturn V's first stage, SpaceX's Falcon 9, Russia's Soyuz, and Atlas rockets all use this fuel combination. Kerosene/oxygen remains popular today because it hits a sweet spot between performance, practicality, and cost.

Cold Fire

Liquid hydrogen and liquid oxygen offer the highest performance of any practical chemical rocket fuel. It powers the upper stages of many launch vehicles. First seriously developed in the 1950s and 1960s, this fuel combination reached maturity with the Saturn V upper stages and Space Shuttle main engines.

The chemistry is simple—hydrogen and oxygen combine to form water, releasing enormous energy in the process. However, both propellants must be kept extremely cold. Oxygen must be chilled to -183°C (90K) and hydrogen to an even more extreme -253°C (20K).

The biggest advantage is efficiency—hydrogen/oxygen engines get more push per pound of fuel than any other chemical rocket fuel. This makes them perfect for upper stages where their high efficiency can really shine. The main drawback is hydrogen's extremely low density, which requires enormous fuel tanks, adding weight and creating design challenges.

The Space Shuttle main engines demonstrated the tremendous power of this fuel combination, though at the cost of complex systems to handle these super-cold liquids. We will talk about how these cryogenic fuels are handled in later chapters.

Cow Burps and Farts

Methane is the newest major player in rocket fuels. While researchers have studied it since the 1930s, it has only recently entered mainstream rocket development with SpaceX's Raptor engine and Blue Origin's BE-4.

When methane burns with oxygen, it produces carbon dioxide and water, along with plenty of energy. It's more efficient than kerosene but less than hydrogen, with better density than hydrogen. This makes methane a "just right" fuel for many rocket applications—not too bulky like hydrogen, not as limited in performance as kerosene, and easier on engine components.

What makes methane very exciting is its potential for Mars missions. Scientists believe we could manufacture methane and oxygen from the Martian atmosphere, allowing spacecraft to refuel for the return trip. And here's me thinking we could have cows do all the methane creation! This makes methane particularly attractive for vehicles designed for Mars missions and reusability, like SpaceX's Starship.

No Matches Needed

Imagine a fuel so reactive it bursts into flame the moment it comes into contact with air—or anything else unlucky enough to be nearby. *Hypergolic* rocket fuels ignite automatically when fuel and oxidizer meet, no spark or flame required. That makes them terrifyingly useful and notoriously dangerous, as John D. Clark noted.

Developed during the Cold War for missiles and later adapted for the Apollo program, these fuels provide extreme reliability. When hydrazine and nitrogen tetroxide meet, they react immediately to produce nitrogen, water, and heat. This instant reaction means these engines can fire thousands of times with no ignition system, making them perfect for spacecraft that need to maneuver repeatedly in space.

The major drawback is that these chemicals are highly toxic and corrosive, requiring special handling procedures and protective equipment. Despite this, their reliability makes them very useful for spacecraft maneuvering systems, attitude control thrusters, and missions requiring long-term propellant storage.

Hypergolic propellants have provided unmatched reliability for critical maneuvers during the Apollo missions, like lunar orbit insertion or landing burns. Their ability to ignite reliably after months in space made them the obvious choice for the Apollo lunar module engines, where failure was not an option.

Hypergolic propellants are specifically valued because they do not need cryogenic storage—they are "storable propellants" kept as liquids at room temperature (or slightly below to ensure safety), unlike many other modern rocket propellants which require deep cryogenic storage.

Beyond Traditional Propellants

The fuels of the future might look nothing like fire and smoke. Rocket propulsion is evolving fast, driven by a mix of daring chemistry and ambitious physics. New "green" propellants offer performance similar to hydrazine without the extreme toxicity. Theoretical propellants like metallic hydrogen could offer extreme performance if scientists can solve practical storage issues.

Nuclear thermal propulsion, which would use a nuclear reactor to heat hydrogen propellant, could potentially double the efficiency of chemical rockets. Even more futuristic concepts like beamed energy propulsion would use lasers or microwaves to heat propellant externally, eliminating the need to carry an energy source onboard the spacecraft.

Each rocket fuel type represents a specific balance of advantages and disadvantages, and mission designers select propellants based on their particular needs. From fire and fuel to light and plasma, the evolution of rocket propulsion is pushing us closer to the stars—and as we'll see in Section 2, the race to perfect those engines is just getting started.

Chapter 9

Getting Into Orbit

Equatorial Slingshot

We live on a giant merry-go-round that is spinning incredibly fast. At its widest point, the equator, we are travelling at 1,670 kilometers per hour (1,040 mph). What happens when you jump off a merry-go-round that spins that fast? You get thrown pretty far off the edge.

This natural motion gives us an amazing opportunity when it comes to launching rockets into space. When rockets get launched from the equator, they too get to use Earth's eastward spin as a free boost, which means they need less fuel to reach orbit, saving millions of dollars per launch. They can either carry heavier satellites and payloads, or use smaller, cheaper rockets to do the same job. The closer you launch to the equator, the bigger this free boost becomes.

That's why launch sites like Kennedy Space Center are tucked into Florida's southern tip—not just for the sunny weather, but for prime positioning closer to the equator. Europe launches satellites from French Guiana to maximize this advantage. Brazil's Alcântara Launch Center, perched nearly on the equator itself, is one of the best spots in the world for takeoff.

This geographic edge shapes the whole space industry, influencing where countries build their spaceports and just how much it costs to get satellites off the ground.

The OG Launch Site: NASA's Wallops Flight Facility (WFF) on Wallops Island in Virginia, United States, is not the most well-known rocket launch site. Established in 1945, it is one of the original launch facilities of NASA and supports science and exploration missions for NASA, federal agencies, and commercial customers. Wallops hosts over a dozen types of launches, including sounding rockets, small orbital rockets, high-altitude balloons, and aeronautical research aircraft. It provides critical launch infrastructure, telemetry, tracking, and range safety services. The facility plays a key role in atmospheric and space science research, technology development, and testing, with a workforce of about 1,100 employees.

Different Orbits, Different Purposes

Not every satellite wants a ride along the equator. TV and internet satellites thrive in equatorial orbits, maximizing the spin boost. But other missions—like weather tracking, global mapping, or reconnaissance—need a bird's-eye view of every spot on Earth, from pole to pole. For these missions, satellites must follow polar orbits—paths that cross over or near the North and South Poles while the Earth spins sideways below then giving them maximum coverage of the Earth surface below.

For launching polar-orbiting satellites, Earth's equatorial spin actually becomes a hindrance rather than a help. When launching into a polar orbit, rockets must actively counteract the eastward momentum they inherit from Earth's rotation. This requires additional fuel and more complex flight paths.

This is why space agencies have developed dedicated launch sites specifically for polar missions:

The Vandenberg Space Force Base in California serves as America's primary west coast launch facility and launched America's first successful polar orbit satellite launch: Discoverer 1 in 1959. Its location allows rockets to launch southward over the open Pacific Ocean, meaning no populated areas lie beneath the flight path.

Newer facilities like Andøya Space Center in Norway take advantage of their high northern latitude. Being closer to the Arctic Circle makes them ideal for launching satellites into polar orbits. The surrounding sparse population and proximity to open ocean create perfect safety conditions.

Scotland's SaxaVord Spaceport in the Shetland Islands offer similar advantages for European missions. These locations combine the benefits of high latitude with direct access to safe drop zones over the North Atlantic, making them increasingly valuable for Earth observation missions.

What Launch Sites Really Are

Modern launch sites aren't just runways for rockets—they're busy hubs built for space activities. Each site is filled with tracking antennas, fuel tanks, advanced safety systems, and teams managing air and sea traffic in real time. The location where a site is built influences nearly everything: international partnerships, the types of rockets used, and which missions can be carried out.

When a country owns its own launch site, it can send missions to space on its own terms. Without one, nations must rely on others for launches—a costly approach that can slow down science and exploration. That's why new players are racing to catch up: Brazil is upgrading its facilities, and the UK is building new launch pads. Meanwhile, private companies are changing the game—creating mobile launchers and even floating rocket platforms that sail out to sea, expanding where and how rockets can launch.

Timing Is Everything

Rocket launches are a careful balance of timing, location, and raw power. Ever thrown a snowball to hit a moving target? You don't just need a strong throw; you need perfect timing. The same principle applies in space: Every launch has to be timed just right, and the site you launch from makes a huge difference.

For missions headed to the International Space Station (ISS), timing is everything. The ISS circles Earth every ninety minutes, so launch teams only have a brief window when the rocket's path aligns with the station's orbit. That launch window might be just a handful of minutes—or just a single second. Miss it, and you wait for the next chance.

Although a rocket must generate enough thrust to break free of gravity and reach the right speed for its destination, even the most powerful rocket won't succeed if its timing and location are off. For some missions—like heading to Mars—the challenge completely changes. Earth and Mars move around the Sun at different speeds and distances, so launches only work when the planets line up. This alignment only happens every twenty-six months. Miss the window and you'll need far more power and fuel—or you might miss Mars completely.

The Invisible Obstacle Course

Golfers in Scotland learn early that the wind at your face rarely matches the wind up in the trees. A ball struck cleanly can rise into a different airflow and drift into another county. That vertical shift is wind shear: abrupt changes in wind speed or direction over height.

Rockets encounter this across hundreds of meters of ascent. The nose may experience a brisk crosswind while the engines enjoy a stiff breeze from an entirely different direction. This creates bending forces that test the rocket's structural limits.

Before launch, teams send weather balloons into the sky, building a vertical map of the atmosphere. If the winds twist too sharply across altitude layers, the countdown pauses. After liftoff, guidance systems monitor the rocket's trajectory and make rapid adjustments—tiny steering corrections that prevent the vehicle from being pushed off its intended path. Engineers may briefly reduce thrust to ease loads, much like slowing a car through standing water to keep traction.

Wind shear influences aviation, energy infrastructure, skyscraper design, and weather prediction. It also explains why driving your tee shot low on a windy day in Troon occasionally feels like a survival skill.

Shedding Weight

Konstantin Tsiolkovsky (1857-1935) was a Russian schoolteacher and physicist who made an important discovery: rockets reach space most efficiently by dropping parts they no longer need. In 1903—the same year the Wright brothers flew—he published *The*

Exploration of Cosmic Space by Means of Reaction Devices, laying out how staged rockets could escape Earth's gravity.

Tsiolkovsky saw the problem with rocket engines hauling empty fuel tanks and spent engines would waste energy. His solution was straightforward—jettison any dead weight along the way. He calculated that stacking multiple stages, each with its own engines and fuel, would allow the rocket to shed mass and get lighter as it climbed. His Tsiolkovsky Rocket Equation showed why this method could propel rockets into orbit, while a single-stage rocket simply couldn't reach the right speed.

Staging is one of the core capabilities of rocket engineering. First stages are built for brute force, designed to lift everything off the ground using engines tuned for sea-level performance. As the rocket rises past the thick atmosphere, upper stages take over, firing more efficiently in space using vacuum-optimized engines with large bell-shaped nozzles.

Real multi-stage rockets appeared in the 1950s. The Soviet R-7 rocket used a central core with four boosters that fell away after burnout, sent Sputnik 1 into orbit and still forms the backbone of Soyuz vehicles, the world's most enduring rocket family. America answered with the Jupiter-C and Explorer 1. The Cold War rivalry rapidly pushed rocket innovation forward. The Soviets launched Vostok—carrying Yuri Gagarin into space—and built powerful rockets like Proton and the giant N1. In the U.S., Mercury-Atlas and Gemini-Titan vehicles pioneered new staging strategies, culminating in the Saturn V. Saturn V, an icon of space travel, stacked three stages: it used the first two to reach orbit and the third to head for the Moon, with performance unmatched for decades. Later, the Space Shuttle introduced parallel staging—its solid boosters and external tanks dropped away as the orbiter continued. Today, commercial rockets optimize staging in new ways: SpaceX's Falcon 9 lands its first stage for reuse, while Ariane 6, China's Long March, Rocket Lab's Electron, and Blue Origin's New Glenn each bring fresh approaches to staging. But the central idea remains: rockets go farther by dropping empty parts.

Hermann Oberth (1894–1989) helped turn theory into reality. Building on Tsiolkovsky's focus on shedding mass, Oberth published detailed blueprints for how humans could reach space, explaining multi-stage rockets, liquid fuel propulsion, and interplanetary trajectories. His 1923 book, *Die Rakete zu den Planetenräumen* (The Rocket into Planetary Space), became a milestone and inspired countless engineers, including Wernher von Braun—the mind behind the Saturn V and Apollo Moon landing.

In the 1930s, Oberth joined von Braun in Germany to advance both scientific and military rocket technology. He lived long enough to witness satellites circling Earth and humans walking on the Moon, fulfilling his earliest visions.

Oberth also invented "hot staging," a technique where the next stage ignites before the previous one detaches, saving (conserving) momentum and optimizing the performance of the rocket. This method appeared first on Soviet rockets and now features in SpaceX's Starship, emphasizing Oberth's key idea that mastering momentum is essential for space travel. Oberth also identified the Oberth effect—when rockets burn fuel while moving fastest, they gain more energy because kinetic energy increases with the square of speed. Maneuvers are most efficient near planets, where the rocket is already moving quickly.

Today's rockets separate their stages using advanced mechanisms—explosive bolts, ullage motors to settle fuels, and engineered systems built to survive brutal shaking, temperature swings, and acceleration. Oberth foresaw these challenges, and his ideas paved the way for the modern, multi-stage rockets that explore our solar system.

The Gravity Turn

Rockets lift straight up punching through the thick lower atmosphere, where air is dense, drag is highest, and the vehicle must carve through a column of increasing pressure. Once they clear this heavy layer, they begin to tilt, allowing gravity to guide the arc. This pitch-over—called the *gravity turn*—sets the rocket on its true path: horizontal motion.

Orbit is not so much about height as it is about speed.

A spacecraft must travel roughly 28,000 kilometers per hour sideways around Earth. At that speed, it continually falls toward the planet yet moves forward fast enough that the surface curves away beneath it. The spacecraft keeps missing the ground. That's orbit: controlled falling–just like Newton's predicted with his falling cannonball idea.

A well-executed gravity turn resembles joining a long motorway slip road. The vehicle shifts direction smoothly and builds speed along the same direction it will ultimately orbit.

Delta-V Budgets

Every time a spacecraft changes speed or direction in space, it uses fuel–this is called "Delta-V," or change in velocity. Delta-V shapes nearly all aspects of spacecraft design: engineers must plan the size of fuel tanks, propulsion systems, and the mission profile for each voyage, all to ensure that the vehicle can achieve every required maneuver.

Keeping a close "Delta-V budget" is essential in missions such as Apollo, which needed enough Delta-V for launch, lunar landing, return, and contingency maneuvers, or for probes like MAVEN and Deep Impact, where miscalculating Delta-V could risk missing the destination or being stranded in space.

To maximize their Delta-V, mission designers often use multi-stage propulsion–each stage drops away when its fuel is spent, making the remaining vehicle lighter and more efficient. This strategy helps reach distant targets but complicates engineering and requires precise timing.

Engineers also take advantage of the Oberth effect–burning fuel when a spacecraft is moving fastest, usually deep within a planet's gravity well, which multiplies the effect of every meter per second spent.

Just like hikers conserve water for emergencies, planners only spend Delta-V when needed, save some for unexpected events, and use natural forces whenever possible. Gravity assists allow spacecraft to gain speed by swinging past planets, giving them a boost without burning fuel.

If the Delta-V budget is underestimated or a maneuver goes wrong, the consequences can be severe: a spacecraft may run out of fuel, miss its target, have to abort its mission, or lose valuable science opportunities. That's why every meter per second is tracked with surgical precision, from escaping Earth's gravity to docking between spacecraft.

Gentle Nudges

Thrust is the force produced by a rocket engine the instant it fires. Impulse is the total effect of that force over time—similar to pushing a child on a swing. You can give one big shove for a quick burst, or many gentle pushes that gradually build up speed. The longer and more you push, the higher the swing goes.

Rockets combine these approaches. The first stage relies on extremely powerful thrust to lift the vehicle off the launch pad. Upper stages, in contrast, use longer and more efficient burns to keep the rocket moving through space. Engineers select engines that balance thrust and duration for each stage of the journey. Just as a careful household manages energy use, upper stages maximize efficiency to conserve fuel.

Impulse is measured in newton-seconds, reflecting both the strength of the engine's push and the time it lasts. This measure is critical for rocket design—a short, powerful burst alone may not be enough to reach orbit, while a weaker engine burning for a longer period can do the job. Rockets achieve their missions by timing engine burns exactly, accumulating enough total impulse to reach orbit, rendezvous with other spacecraft, or travel between planets.

The Bottom Line

Reaching orbit requires meticulous mastery of location, timing, weight, and velocity. Geography provides boosts. The atmosphere provides hazards. Delta-V sets the boundaries of what can be done throughout a mission. Staging manages mass. Guidance systems thread through the wind. And engines convert burning propellant into the steady accumulation of impulse that lifts the vehicle beyond the sky.

Getting into space is such a multi-disciplinary adventure of incredible physics and engineering. Staying up there is a whole other set of disciplines.

Chapter 10

Orbital Mechanics

Flying in Space

Orbital mechanics is the science of how things move under gravity—why planets stay in their paths, why satellites circle Earth, and how spacecraft travel across the solar system. It lets us plan missions so rockets, probes, and satellites depart, swing by, and arrive exactly where and when intended. And it is all built upon the important concepts and laws and Newton brought us.

The Physics of Staying Up

A satellite in orbit isn't actually hovering in place—it's constantly falling toward Earth. But it's moving sideways so fast (about 7.8 km/s) that Earth's surface curves away at the same rate. It's a perpetual freefall that never reaches the ground.

While circular orbits are a special case, most orbits are elliptical—oval-shaped paths with two key points: the perigee, the closest point to Earth where the satellite moves fastest, and the apogee, the farthest point where it moves slowest.

Orbits become elliptical because of how gravity and the satellite's forward momentum interact. When a satellite moves closer to Earth, gravity pulls it in stronger, causing it to speed up. As it moves farther away, gravity weakens, and it slows down. This change in speed, combined with the satellite's momentum, stretches the path into an ellipse rather than a perfect circle.

This balance of forces and energy conservation is what Johannes Kepler described in his famous laws of planetary motion. Unlike circular orbits where speed and distance remain constant, elliptical orbits naturally produce variations in speed depending on where the satellite is on its path—faster near perigee, slower near apogee.

Today's orbital mechanics mixes these centuries-old laws of motion with modern computing, allowing us to plot and adjust satellite paths with incredible precision. Whether it's the International Space Station (ISS) buzzing overhead every ninety minutes or communication satellites locked in their perfect geostationary spots, all obey the same elliptical laws Newton and Kepler uncovered. The mathematics have not changed—it's just that now, we know how to put it to work better than ever.

Orbital Planes

Orbital planes are like unseen highways in space that satellites follow as they circle Earth. Think of slicing through the planet with a giant sheet of glass, with the path a satellite takes lying flat on that sheet. While this idea seems simple today, it was once a major breakthrough in understanding how objects move through space.

Johannes Kepler was among the first to realize that orbits aren't random but lie within distinct planes. In the early 1600s, while studying planetary motion, he suggested that all the planets orbit the Sun roughly in the same flat band—the *ecliptic plane*. This was revolutionary because it revealed a new hidden order beneath what had seemed like chaotic celestial movements.

Today, we select orbital planes to suit different missions. For example, GPS satellites are arranged in six separate orbital planes, spread out like segments of an orange around Earth. This ensures that wherever you are on the planet, there are always enough satellites overhead to provide an accurate location fix.

The angle between an orbital plane and Earth's equator is called the *inclination*. A satellite orbiting directly over the equator has 0° inclination, while one going over the North and South Poles has 90° inclination. Choosing the inclination determines what parts of Earth a satellite can see and how well it can perform its mission.

Orbital planes are like train tracks in space–they keep satellites moving on predictable routes. This helps us manage all the spacecraft circling our planet, whether they're connecting our phones, watching the weather, or studying the stars.

Orbital Decay

Operating a satellite isn't as simple as launching it into space and then forgetting about it. Satellites face a variety of challenges that can gradually push them off course–but we have ways to keep them on track.

Earth itself creates the first problem. Our planet has an uneven shape, which means gravity pulls differently in different places. These uneven pulls can slowly knock satellites off course. To fix this, satellites have small rocket engines that fire every now and then to nudge them back into the right position–similar to how you'd make small adjustments to your steering wheel to keep your car in its lane.

But Earth's gravity isn't the only force at play. The Moon and the Sun also exert gravitational tugs, and even sunlight itself applies some pressure. Satellite operators constantly monitor these little nudges using sophisticated computer models and plan regular adjustments to keep satellites steady–like a boat captain who's always checking weather forecasts and adjusting the rudder to keep the ship sailing smoothly.

Satellites closer to Earth have yet another challenge: the thin remnants of our atmosphere create drag, slowing them down and causing their orbits to decay. To counter this, these satellites carry extra fuel to perform periodic "boost burns"–brief thruster firings that propel them back to the correct altitude, much like giving a weary runner a quick energy boost to help them keep pace.

Thanks to these constant, careful corrections, satellites can maintain their exact orbits, continuing to serve us reliably day after day from hundreds of kilometers above.

Orbital Neighborhoods

Space around Earth is structured into regions, each with its own character and purpose. Together, these orbital layers support nearly every modern technology we use today.

The Busy Neighborhood

Low Earth Orbit (LEO), stretching from about 160 to 2,000 kilometers up, is the bustling ground floor of space. Satellites here whip around the planet in about 90 minutes at close to 28,000 kilometers per hour, circling Earth roughly sixteen times during one sidereal day—the twenty-three hours, fifty-six minutes it takes Earth to complete a full rotation relative to the stars. At that speed, an object could cover the distance from Glasgow to Sydney in well under an hour, if air travel worked like orbital motion.

This region hosts the International Space Station, Earth-observation satellites, reconnaissance platforms, and the surging constellations of communication satellites such as Starlink. LEO's proximity to Earth ensures high-resolution imaging and minimal signal delay, which makes it ideal for everything from climate monitoring to global broadband.

It also hosts the debris problem. Discarded rocket parts, defunct satellites, and fragments from past collisions drift through LEO at speeds where even a paint fleck carries a heck of a punch. In 2009, the collision between an active Iridium satellite and a dead Russian craft scattered thousands of pieces across LEO, each one a hazard to anything crossing its path. Tracking networks now monitor tens of thousands of objects, and satellites perform regular avoidance maneuvers to maintain safety.

Polar Bears

Within the LEO altitude range, there are two important types of orbits: polar and Sun-synchronous.

Polar orbits pass over (or very close to) both poles with each revolution. As Earth turns beneath them, these satellites gradually scan the entire planet, making them vital for mapping, reconnaissance, and weather forecasting.

Sun-synchronous orbits take a different approach. By selecting the right altitude–usually between 600 and 800 kilometers–and tilting the orbit around 97 to 98 degrees from the equator, engineers use Earth's equatorial bulge. The planet's uneven mass distribution causes the orbital plane to slowly precess around Earth's axis. When set at the correct inclination, this precession matches Earth's yearly orbit around the Sun, ensuring the satellite passes over each region at the same local solar time. This steady sunlight angle helps scientists compare images taken months or even years apart, being confident that any changes observed are due to Earth, not lighting conditions. Vegetation changes, glacier retreats, city growth–all become clear thanks to this orbital stability.

The Navigation Highway

Medium Earth Orbit (MEO), from 2,000 to 35,786 kilometers, is largely the domain of timing. GPS, Galileo, GLONASS, and BeiDou satellites live here, orbiting at around 20,200 kilometers in carefully arranged constellations.

At this altitude, each satellite completes two orbits per sidereal day, creating the geometric regularity needed for highly precise navigation. Atomic clocks aboard each satellite keep time to billionths of a second, although relativity requires correction. Being farther from Earth's gravity makes the clocks tick faster; travelling at high orbital speeds slows them slightly. Ground teams account for both effects to maintain the accuracy upon which modern society relies.

These timing layers form the backbone of global navigation: planes, ships, financial markets, telecom networks, emergency services–all depend on signals from satellites calmly circling far above.

Hovering Dots

Arthur C. Clarke imagined it first: a satellite positioned so far above the equator–35,786 kilometers–that its orbital period matched Earth's rotation. At that height, the satellite appears fixed in the sky, never drifting left or right. Antennas on the ground can point once and forget.

Syncom 3, launched in 1964, turned Clarke's concept into working machinery. Today, this band over the equator hosts weather monitors, broadcast satellites, secure communications, and long-haul data networks. It is the penthouse suite of orbital real estate.

Reaching Geostationary Earth Orbit (GEO) involves a well-planned Hohmann transfer, lifting the satellite from low Earth orbit into an elongated elliptical path before smoothly circularizing it at GEO altitude, approximately 35,786 kilometers above Earth's equator. Once in position, the satellite appears fixed over a single longitude, needing only small station-keeping burns to counter solar radiation pressure and the subtle gravitational tugs from Earth's uneven mass and the pull of the Sun and Moon.

While geosynchronous orbits share the same twenty-four-hour period as Earth's rotation, they don't always have to sit directly above the equator. If the orbit is inclined or slightly elliptical, the satellite appears to trace a slow figure-eight pattern, or analemma, in the sky over the course of a day, drifting slightly north–south and east–west around a central point. These inclined geosynchronous orbits are especially helpful for reaching higher-latitude regions, where a perfectly equatorial geostationary satellite would see at a low angle, thus improving coverage and signal quality for users far from the equatorial belt.

Lingering Orbits

Beyond GEO lies a realm of highly specialized paths. The Molniya Orbit, devised by Soviet engineers in the 1960s, solved the challenge of providing reliable coverage to northern latitudes. This elongated ellipse brings the satellite close to Earth at one end and high above the northern hemisphere for most of its twelve-hour orbit.

The secret lies in a tilt of 63.4 degrees, which cancels natural orbital precession caused by Earth's bulge. With three satellites spaced 120 degrees apart, continuous northern coverage becomes possible. The design remains elegant—and in use—decades later.

Future Challenges and Opportunities

Space has grown busy. LEO brims with satellites and fragments; MEO and GEO are filling; and plans for vast next-generation constellations surge ahead. These expanding orbital neighborhoods demand careful stewardship.

Tracking networks now serve as an informal air traffic control system for space, continuously cataloging objects in orbit, predicting close approaches, and alerting operators so they can perform avoidance maneuvers. Engineers are developing debris-removal tools–like capture nets, robotic arms, drag sails, and targeted deorbit vehicles–to grab or slow down space junk, helping it safely burn up in the atmosphere instead of becoming dangerous shrapnel. At the same time, regulators are putting rules in place that encourage satellite operators to design their spacecraft with disposal in mind, requiring them to deorbit or move to designated graveyard orbits at the end of their missions–ensuring space stays safe and accessible for everyone.

Farther from Earth, orbital mechanics shapes our ambitions across the solar system. Gravity assists offer free boosts by stealing a little momentum from planets. Lagrange points–areas where gravitational forces balance–provide ideal parking spots for observatories like the James Webb Space Telescope.

Every weather report, GPS fix, video call, and satellite broadcast flows through this hidden architecture. Orbital mechanics, therefore, is critical in shaping our modern civilization with incredible precision.

But mastering how to stay in orbit is only the beginning. To explore the solar system, we must also learn how to move through it.

Chapter 11

Intergalactic Highways

Moving Through a Moving Solar System

In the previous chapter, we saw how satellites stay aloft and how gravity shapes their paths. Now, the same laws must be used to leave Earth behind.

Reaching Earth orbit is just our first step into space. Once a spacecraft reaches orbit, it's ready to venture toward other worlds in our solar system. This journey requires careful planning since planets are constantly moving, each orbiting the Sun at its own speed. Everything is flying about!

Remember how Kepler found that planets move in elliptical orbits around the Sun, while Newton explained how gravity works to control these movements? Well, space travel follows those natural laws of Kepler and Newton.

When planning a space mission, engineers must consider many moving parts. They track the changing positions of planets, moons, and other objects, each with its own gravitational pull. The spacecraft's path must weave through these gravitational fields, and engineers need to calculate exactly when to fire the engines.

The Hohmann Transfer

Back in 1925, long before anyone launched a spacecraft or orbited Earth, German engineer Walter Hohmann described a key principle still used to navigate the solar system. As both an architect by trade and a visionary, Hohmann published his landmark book, *Die Erreichbarkeit der Himmelskörper* (The Attainability of Celestial

Bodies), explaining how to move a spacecraft efficiently between two orbits. This is what we now refer to as the Hohmann Transfer Orbit.

Hohmann's idea was a groundbreaking step in orbital mechanics long before satellites started orbiting Earth. You can think of a Hohmann transfer like an on-ramp onto a new orbital highway: it helps a spacecraft switch from one nearly circular orbit to another with just the right amount of fuel. The maneuver uses two well-timed engine burns. The first one nudges the spacecraft off its initial orbit onto an elliptical transfer path. After drifting along this path toward the target orbit, a second burn adjusts its speed so it can seamlessly merge into its new orbital lane, avoiding overshooting or drifting away.

Picture changing lanes on a vast motorway that stretches across hundreds of millions of kilometers. The timing and speed are crucial—if either burn happens too early or too late, the spacecraft might miss its target or end up using more fuel than necessary. That's why Hohmann's method is so popular: it's an efficient way to move between orbits, saving both weight and cost, making it a smart choice for space missions.

NASA's Mariner 2 was the first spacecraft to use a Hohmannstyle transfer to reach another planet, arriving at Venus in 1962. Since then, most missions to Mars, Mercury, and the outer planets have started with some form of Hohmann transfer. Even when modern spacecraft add gravity assists or low-energy routes for extra efficiency, Hohmann's insight remains the backbone of interplanetary mission design.

Gravitational Highways

Have you ever watched a leaf drifting along a calm river, carried smoothly by the current? In space, something similar is possible. Spacecraft can navigate natural gravitational forces along low-energy paths which act like secret highways through the solar system, allowing them to travel huge distances while using very little fuel.

For missions that do not need to arrive quickly, the solar system offers this slower but thriftier option. In the late 1980s, Edward Belbruno, working at NASA's Jet Propulsion Laboratory, showed that gravity carves subtle channels through space–low-energy pathways that connect regions around planets and moons. Together these channels form what is now called the Interplanetary Transport Network (ITN), a kind of cosmic rail system where gravity does most of the work.

Moving along these gravitational corridors feels like drifting down a lazy river. A spacecraft glides from one gravitational basin to the next with only tiny steering burns, hopping between currents while burning very little propellant. The journey can take a long time, but the fuel savings allow heavier instruments, longer lifetimes, and mission ideas that would otherwise be impossible.

Japan's Hiten spacecraft proved the value of this approach. Running low on fuel and unable to reach the Moon with conventional burns, mission planners placed it on a low-energy drift through the Earth-Moon gravitational field; after a patient cruise, it arrived in lunar orbit with propellant to spare. NASA's Genesis mission used similar strategies, travelling more than thirty-two million kilometers while using remarkably little fuel.

Imagine space travel options like choosing between a speedy express train and a more leisurely freight train. The express train–the Hohmann transfer–gets you there quickly and straight to the point, but it uses up more fuel–kind of like shoveling extra coal into the fire. On the other hand, the freight train–the ITN–takes a bit more meandering, using less energy but taking longer to arrive. Both methods are important; a balanced transportation system benefits from having both fast services and slower, more efficient routes.

The Cosmic Slingshot

While low-energy transfers rely on drifting through gravitational currents, another tool in space navigation is the *gravity assist*, or slingshot. Instead of burning through fuel to pick up speed, a spacecraft can tap into a planet's own motion, using its gravity to fling itself forward like a rock in a sling. This idea was first introduced in the 1920s by Soviet scientist Yuri Kondratyuk and finally became reality decades later.

It is a bit like walking through a busy airport and stepping onto a moving walkway while already in stride. Your own momentum combines with the walkway's motion, and suddenly you glide forward far faster than your legs alone could manage. Just time your exit well!

A gravity assist in space follows the same principle. When a spacecraft approaches a planet along the planet's direction of travel and at just the right angle, the spacecraft draws on the planet's orbital momentum. The result is a powerful, fuel-free surge in speed that sends the craft outward with impressive efficiency.

The timing and geometry of the flyby control how big that boost will be, so mission designers choose paths that deliver the best tradeoff between speed and destination. Gravity provides the pull; the planet's motion provides the push; the spacecraft departs with more energy than it arrived with.

This is a very good example of Newton's Third Law: the planet tugs on the spacecraft, and the spacecraft tugs back, but the planet is so massive that its motion barely changes, while the spacecraft ricochets away with a big change in speed.

The first major test of gravity assists came in 1974, when NASA's Mariner 10 used Venus as a slingshot on its way to Mercury. The Voyager missions then showed how powerful the technique could be. Launched in 1977, Voyagers 1 and 2 took advantage of a rare alignment of the outer planets, flying past Jupiter, Saturn, Uranus, and Neptune in succession and using each world as a slingshot toward the next. Voyager 1's close pass by Jupiter in 1979 increased its speed by about a third, with extra boosts from Saturn, while Voyager 2 went on a grand tour of four giant planets over twelve years. By the end, both spacecraft were racing out of the solar system at more than seventeen kilometers per second—fast enough to escape the Sun's gravity entirely.

Gravity assists helped Voyager 1 become the first humanmade object to enter interstellar space in 2012, followed by Voyager 2 in 2018. Today, Voyager 1 is tens of billions of kilometers from Earth, still coasting outward on that carefully planned sequence of planetary flybys—a perfect example of Newtonian physics doing the heavy lifting

Modern Missions

Today's missions combine these techniques in a way that would make Hohmann and Kondratyuk proud. ESA's BepiColombo uses flybys of Earth, Venus, and Mercury to slow itself enough to orbit the innermost planet, a task made difficult by the Sun's enormous gravitational pull. NASA's Parker Solar Probe performs repeated passes of Venus to tighten its loop around the Sun, diving closer to the star than any spacecraft before it.

Juno stole a speed boost from Earth to reach Jupiter. Cassini used the moons of Saturn as stepping stones, adjusting its path with each pass. ESA's Rosetta spiraled through a series of Earth and Mars flybys before catching a comet. New Horizons picked up a decisive shove from Jupiter to make its historic dash past Pluto.

SpaceX plans to blend these techniques with orbital refueling to make space travel more efficient. Starship will launch to Earth's orbit, then meeting up with tanker vehicles to top up its fuel. During the favorable Earth-Mars alignment that happens roughly every twenty-six months, multiple ships can head out, gradually building up supplies on Mars with steady cargo deliveries. Though a Hohmann-style transfer might not be the fastest method, it's excellent for maximizing payload capacity–the key to establishing a long-term presence.

No matter the mission, the idea remains the same: work with gravity whenever possible, only oppose it when necessary, plan your route carefully, and let the solar system's natural forces do as much of the heavy lifting as physics allows.

Parking Spaces

What if there were places in space where a spacecraft could just "park" and stay put without constantly using fuel to keep its position? At these special places, the gravity from two huge objects, such as the Earth and the Sun, balances out perfectly. Understanding this concept helps us find the best spots to position satellites and space missions, saving fuel and making exploration easier.

The idea dates back to 1772, when Italian-born French mathematician Joseph-Louis Lagrange (1736-1813) shared a fascinating insight into gravity and motion. He showed that when you have two massive objects orbiting each other—such as Earth orbiting the Sun—their combined gravitational forces create five special locations in space. These spots, known as the Lagrange Points L1 through L5, allow smaller objects like spacecraft or even asteroids to remain "parked" with minimal effort.

To visualize this, imagine rolling a marble across a twisted, curved surface. It naturally settles into a low spot and stays there, balanced by gravity. Lagrange Points work in a similar way, but in three-dimensional space. At each of these points, three forces come into perfect balance: the gravitational pull from Earth, the gravitational pull from the Sun, and a centrifugal force created by the object's own motion as it orbits the Sun (similar to the sideways push you feel when a car takes a sharp turn). When these forces align just right, they form stable or semi-stable "pockets" where a spacecraft can effectively hover, using little to no fuel to maintain its position.

Think of a ball resting perfectly still on a flat spot on a hill—neither rolling downhill nor tumbling off. These balance points allow spacecraft to remain in place efficiently, making them extremely valuable for long-term missions that require a stable observation or communication platform far from Earth.

L1: The Sun Watcher

Located between Earth and the Sun at about 1.5 million kilometers from our planet, L1 serves as a critical vantage point for solar observation spacecraft. At this position, the combined gravitational forces of the Earth and Sun create a relatively stable environment where spacecraft can maintain their position while using minimal fuel. The SOHO (Solar and Heliospheric Observatory) spacecraft has been operating here since 1995, providing continuous monitoring of the Sun's activity and capturing incredible images of solar flares and coronal mass ejections. It acts as an early warning system, giving us fifteen- to sixty-minute advance notice of potentially hazardous solar storms heading toward Earth. NASA's Advanced Composition Explorer (ACE) and Deep Space Climate Observatory (DSCOVR) satellites also occupy this strategic location, working together to provide real-time measurements of solar wind conditions and detailed data about space weather phenomena that could impact our telecommunications, power grids, and satellite operations.

L2: The Deep Space Observatory

Located approximately 1.5 million kilometers (about a million miles) behind Earth from the Sun's perspective, L2 is a really important location for space observation. The James Webb Space Telescope (JWST) operates from this strategic point because it offers unique advantages for deep space observation. At L2, the telescope can maintain a fixed position where Earth, the Sun, and the Moon all remain on the same side, creating a stable thermal environment. This alignment allows JWST's enormous sunshield to block light and heat from all three bodies simultaneously, keeping its sensitive infrared instruments at the extremely cold temperatures (below -233°C or -388°F) required for optimal performance.

The position also provides an unobstructed view of deep space for about half of the sky at any given time, and the full sky over six months. L2's relative stability also means the telescope requires minimal fuel for station-keeping maneuvers. The European Space Agency's Gaia spacecraft also takes advantage of these favorable conditions at L2, where it's creating the most detailed three-dimensional map of the Milky Way ever attempted, measuring the positions, distances, and motions of billions of stars with unprecedented clarity.

China's Queqiao relay satellite, operating from L2, helps us learn more about the Moon's far side. By studying lunar geology through this satellite, scientists can better understand how Earth formed and evolved, potentially revealing more about our planet's natural resources and history.

L3: Hidden by the Sun

L3 sits directly on the opposite side of the Sun from Earth, approximately 150 million kilometers away. This point maintains the same orbital period as Earth, always remaining hidden behind the Sun. While it might seem like an ideal location for observing the Sun's far side or monitoring solar activity, no spacecraft have been stationed there. The main obstacle is communications—radio signals would need to travel through or around the Sun to reach Earth, making reliable contact nearly impossible with current technology. The Sun's intense radiation and interference would also likely disrupt any communication attempts. Despite these limitations, L3 remains theoretically important for understanding orbital mechanics and the complex interplay of gravitational forces in our solar system. It also serves as a useful example of how some mathematically stable points may be impractical for real-world space operations.

L4 and L5: The Stable Twins

These points, known as L4 and L5, follow Earth's orbital path around the Sun in unique positions. L4 travels sixty degrees ahead of Earth, while L5 follows sixty degrees behind. Together with Earth and the Sun, they form two equilateral triangles in space and objects placed there tend to stay put naturally. This happens because the gravitational pulls from both Earth and the Sun combine to keep objects balanced, much like a ball resting in a valley that keeps it steady instead of rolling away.

The stability of L4 and L5 points is dramatically demonstrated in our solar system by Jupiter's Trojan asteroids. Over 12,000 asteroids have been discovered in Jupiter's L4 and L5 points, with scientists estimating there may be as many as 1.6 million objects larger than 1 kilometer in diameter trapped there. These Trojans have been orbiting at these points for billions of years. The asteroids are divided into two "swarms"—the Greek camp at L4 and the Trojan camp at L5, named after opposing warriors in Homer's *Iliad*.

Future missions will treat these points as waystations: observatories, communications hubs, early-warning centers, and perhaps one day fuel depots. In a solar system with vast distances and slow travel, these gravitational pockets will serve as stepping stones.

Highways and Pathways

The deeper we map these orbital pathways, the more the Solar System reveals itself as a connected system rather than just a large empty void. This network of highways offers us many ways to move around depending on our needs.

Hohmann transfers give us the efficiency of patient timing, while gravity assists harness the free energy from entire worlds. Lagrange points offer pockets of remarkable stability in space, and the Interplanetary Transport Network forms a network of low-energy corridors that link planets, moons, and asteroids.

Each highway and all of the explorers we have sent out into the solar system shows us that gravity, geometry, energy, and our own ingenuity can create a navigable geography and adventure our predecessors could only have imagined.

All of our old friends, from Kepler to Newtown, Boyle to Bernoulli, and Euler to Einstein, have laid out the system of how everything moves.

Now, we need to make the first step towards those sparkly things in the sky: getting off the ground and into orbit.

And that's Science Class over for now.

SECTION 2: LIFTING OFF

From Fireworks to the Final Frontier

We've spent time with thinkers, dreamers, scientists, and astronomers–the ones who mapped the heavens, timed the tides, showed us that motion has rules, and helped us understand how gas flows and temperature and pressure behave. Now we turn to the pyromaniacs who actually set fire to things and built machines that defy gravity.

To break free from Earth's gravitational prison, you need a *lot* of power. Transforming molecular bonds into fire-breathing machines and accelerating large structures through the atmosphere at hypersonic speeds –often exceeding Mach 25 (19,000 mph)–means containing ferocious chemical reactions, violently expanding superheated gases, and structures that can handle cryogenic temperatures, and that do not melt under extreme temperatures close to that of the sun.

Today's launch vehicles are multi-stage machines of power, with special alloys and complex computer systems, each built to carry payloads and people beyond the sky.

This section is the story of how that happened. Of how fire turned into flight. Of how trial, error, and relentless engineering carved a path from smoke-filled fields to the vacuum of space.

We're not yet ready to unpack the full rocket science behind combustion chambers, supersonic nozzles, or orbital transfers. That will come.

But first, we light the fuse and step back–to see where it all began.

Chapter 12

Flying Fire

————————◆·◆·◆————————

From Fire Arrows to Warfare Rockets

Long before rockets took astronauts into space, they were actually fierce and unpredictable tools that caused chaos and fear on medieval battlefields. Even before modern space exploration, these early devices mainly served to terrorize enemies with their wild and unpredictable fury.

Back in ninth-century China, Taoist alchemists seeking immortality made an ironic discovery: gunpowder. This surprisingly explosive mix of saltpeter, charcoal, and sulfur became a versatile invention. Over time, they turned it from spectacular fireworks into powerful fire lances, and eventually into the groundbreaking "fire arrows" of the Song Dynasty–really the first real rockets. Funny to think that the pursuit of eternal life unexpectedly sparked a breakthrough that transformed warfare and technology.

These primitive rockets were made from bamboo tubes packed with powder, attached to wooden shafts. Although they were not very accurate, they didn't need to be–their power lay in spectacle, filling the air with smoke, noise, and unpredictability.

By the early thirteenth century, fire arrows became standard in Chinese warfare. During the 1232 siege of Kaifeng, defenders launched "flying fire" against Mongol forces, who then adopted and spread the technology westward through Asia, the Islamic world, and eventually to Europe.

Chinese engineers continued refining their designs through trial and error. Despite frequent failures–rockets that sputtered, exploded prematurely, or veered backward—each attempt improved their understanding: controlled combustion generates thrust, and thrust creates distance.

Despite being dangerous and unreliable, the concept endured: fire could become force. Newton hadn't yet formalized his laws of motion, but Chinese rocketeers were already demonstrating them empirically. Humans were now using stored chemical energy to propel objects across distances without physical contact—the conceptual ancestor of every modern rocket.

The Rise of Mysorean Rockets

By the eighteenth century in southern India, the Kingdom of Mysore made real advances in rocketry, turning it from a sparkly novelty into a powerful tool on the battlefield, especially under leaders like Hyder Ali and Tipu Sultan.

Their success was partly thanks to innovative metallurgy: they replaced bamboo with hammered iron casings that could handle more pressure, hold larger amounts of gunpowder, and produce greater thrust. These small iron cylinders, mounted on bamboo sticks, sometimes even carried blades, transforming into unpredictable aerial weapons aimed at intimidating their enemies and turning the tide of warfare.

In 1780, at the Battle of Pollilur, Mysorean rocket brigades devastated British East India Company forces with coordinated barrages. The sky lit with fire trails as British supply wagons exploded and formations collapsed—a rare instance where indigenous technology decisively outclassed European forces.

Tipu Sultan took a significant step forward by establishing dedicated rocket corps, consisting of around 5,000 rocketeers who worked together as a well-organized artillery team. The ingenious Mysorean rocket showcased that metal casings could withstand higher pressures, and with careful design, stable flight was achievable. This progress moved rocketry from being just a spectacle to becoming a more controllable and powerful weapon.

The Potential of the Congreve Rocket

As the nineteenth century approached, Britain had won the Anglo-Mysore Wars but was profoundly affected by Tipu Sultan's iron-cased rockets. Sir William Congreve (1772-1828), whose father was the Royal Arsenal's Comptroller, saw an opportunity to engineer his own version.

Starting in the early 1800s, Congreve had developed rockets that could fly up to two miles with more consistent thrust and improved stability. Rather than aiming for pinpoint accuracy, they were designed for psychological impact—overwhelming fortifications, setting ships ablaze, and scattering infantry with thunder and flame.

These rockets played an important role during the Napoleonic Wars and are best remembered for their part in the 1812 bombardment of Fort McHenry. It was there that Francis Scott Key captured the scene of "the rockets' red glare," which later inspired the words of the US national anthem.

Despite some limitations—like inaccuracy, susceptibility to weather, and transport difficulties—Congreve rockets were the first government-backed rocketry program that became part of military strategy. They had proven that rockets could be manufactured in large quantities and engineered for predictable effect.

By the 1850s, Congreve rockets had fallen out of regular use, replaced by rifled artillery on the battlefield and mainly reserved for ceremonies. Yet, they represented the beginning of an exciting new chapter. Congreve's lasting contribution was pioneering a standardized, mass-produced projectile that didn't need a launcher or line of sight, and that stored chemical energy in a metal casing. This demonstrated that chemical propulsion could be controlled, manufactured, and directed. From Congreve's arsenals to today's launchpads, the fundamental idea remains the same: thrust equals power.

British inventor William Hale (1797-1870) improved on this foundation by removing the guidance stick and introducing spin-stabilization. This made rockets more accurate and respectable while suggesting a new question: what could these machines accomplish beyond warfare? What if rockets weren't instruments of destruction but invitations to explore?

Theological Rockets

In 1861, the Scottish astronomer and theologian William Leitch published an essay, *A Journey Through Space*, a work that reads today like a preface to modern astronautics. Reflecting on how a vehicle might move beyond Earth, he wrote that "the only machine, independent of the atmosphere, we can conceive of, would be one on the principle of the rocket." He explained that a rocket rises "not from the resistance offered by the atmosphere to its fiery stream, but from the internal reaction," adding that its velocity "would, indeed, be greater in a vacuum than in the atmosphere." In other words, he understood that propulsion in space does not require pushing against air at all. Decades before Tsiolkovsky formalized the rocket equation, Leitch had already grasped the essential physics: Newton's laws, reaction mass, and the vacuum of space working together as a natural trio.

Modern Rocketry

At the end of the 19th century, rockets were simply fireworks and decorative displays for celebrations. But in a quiet corner of Russia, their exciting future was being imagined and planned right at a humble kitchen table.

In a previous chapter, we met Konstantin Tsiolkovsky and explored his Rocket Equation—the amazing formula that explains how rockets work. His equation reveals how rockets move forward by pushing mass out behind them. The speed of a rocket depends mainly on two things: how quickly the exhaust is expelled and the difference between the rocket's initial weight and its final weight.

This relationship is still so important today because it influences every rocket launch. It tells us how much cargo a rocket can carry and how much fuel it needs. The equation also helps us understand why rockets drop stages as they ascend—discarding each empty section makes the rocket lighter and allows it to accelerate faster.

Even a small reduction in weight can dramatically boost a rocket's performance. Imagine hiking with a backpack—removing just one pound can make walking much easier. For rockets, the weight you first remove makes the biggest difference, and each subsequent reduction contributes less and less. That's why rocket engineers are so dedicated to saving every little gram they can.

The first real tests of multi-stage rockets took place during World War II in Germany. The V-2 rocket had only one stage, but German engineers began exploring designs with multiple stages. The Wasserfall anti-aircraft missile incorporated some early ideas for two-stage design, although it never actually flew in combat before the war ended.

After World War II, both the United States and the Soviet Union acquired German rocket technology and experts. By the late 1940s, they started testing early multi-stage rockets. The American "Bumper" program, for example, combined a V-2 rocket as the first stage with a smaller WAC Corporal on top as the second stage. On February 24, 1949, it reached an altitude of 393 kilometers (244 miles)—marking the first time a US rocket went into outer space and launching the first successful multi-stage high-altitude flight.

These early tests showed that Tsiolkovsky's ideas were right: using stages made rockets work better and go farther. Although these rockets were still just experiments, not practical vehicles, they were important steps forward.

Tsiolkovsky also dreamed of a future where space was part of our world. He imagined space stations with greenhouses, colonies on the Moon, and missions to Mars—long before most people believed such things were possible, at a time when steam engines still powered most of the world. He suggested that rockets should use liquid fuels like hydrogen and oxygen instead of gunpowder, which was a bold idea for his time. He even described features of space stations, like systems to recycle air, pressurized cabins to keep astronauts safe, and spinning sections to create artificial gravity—all concepts that wouldn't become reality for many decades.

Most people back then saw Tsiolkovsky as just a rural schoolteacher with wild science fiction dreams. But he kept working tirelessly. Years later, Soviet engineers built on his ideas to create their entire space program. Sergey Korolev (1907–1966) specifically relied on Tsiolkovsky's equations and ideas when developing Sputnik and planning Yuri Gagarin's historic flight.

Tsiolkovsky famously said, "Earth is the cradle of humanity, but one cannot remain in the cradle forever." He believed that humanity was meant to become a civilization among the stars. Working alone in rural Russia, with little money or recognition, Tsiolkovsky laid the first steps on our journey to space and changed how people think about rockets—transforming them from weapons into powerful tools for exploration.

The Godfather of Rockets

If Tsiolkovsky was the philosopher of space, and Oberth helped us figure out how to reach orbit, Robert H. Goddard (1882–1945) was the hands-on engineer, working with greasy tools and safety goggles. Born in Worcester, Massachusetts, Goddard combined science fiction dreams with real physics. After reading H.G. Wells as a teenager, he dedicated his life to making spaceflight a reality.

Unlike Tsiolkovsky, who worked with pencil and paper, Goddard turned his backyard into a workshop and snowy fields into test sites. On March 16, 1926, he launched a ten-foot metal cylinder that flew for just 2.5 seconds and reached only forty-one feet high—but it was the world's first liquid-fueled rocket flight.

Goddard's innovations played a crucial role in turning rockets from wild, unpredictable devices into reliable, controllable machines. His contributions laid the groundwork for modern rocketry. He developed gyroscopic stabilization systems—spinning wheels that could sense when a rocket started tilting and help keep it pointing in the right direction, similar to how your smartphone knows which way is up. He also invented thrust vectoring by creating rocket nozzles that could move and redirect the exhaust—like pointing a garden hose in different directions—allowing rockets to steer while flying.

Goddard also built advanced fuel injection systems instead of merely dumping fuel into the engine. These systems managed how fuel and oxidizer mixed together, much like the systems in modern cars that regulate gasoline flow. To handle the extreme heat that could melt rocket engines, he designed cooling systems that pumped fuel around the hot parts before burning it, absorbing dangerous heat.

These inventions continue to form the backbone of all modern rockets that send satellites, astronauts, and space probes into space today. Back in 1920, newspapers ridiculed Goddard for saying rockets could operate in space. They mistakenly believed rockets required air to push against. It wasn't until forty-nine years later, when Apollo 11 was on its way to the Moon, that *The New York Times* finally acknowledged they were wrong.

With limited support and mostly private funding from the Guggenheim family, Goddard persisted in conducting his research in a humble barn-like laboratory. The US military didn't pay much attention to him, but German engineers took a keen interest in his patents. After World War II, it was discovered that Wernher von Braun (1912-1977) had incorporated many of Goddard's innovations into his German rocket projects.

Goddard passed away before anyone reached orbit, but his work revolutionized rocket design. That's why NASA named the Goddard Space Flight Center in his honor.

From simple bamboo tubes to advanced liquid-fueled engines, rockets have carried our hopes, our fears, and our dreams. Each ignition—whether on a battlefield or in a snowy Massachusetts field—brought us closer to exploring the skies. The fire that once aimed to destroy now drives us to discover.

Chapter 13

Space Races

Supersonic Stuff

By the late 1930s, rockets really took a giant leap from science fiction into real-world conflicts. What once seemed destined for moon landings quickly became a key part of warfare. During World War II, the pace of rocket innovation sped up dramatically, blending the power of military might with dreams of exploring space.

The German V-2 rocket was a major milestone. It stood fourteen meters tall and was fueled by liquid oxygen and ethanol, becoming the first supersonic missile and the first human-made object to reach the edge of space. It flew as high as eighty-eight kilometers (and could reach 206 kilometers if launched straight up), forever expanding what rockets could do.

Much of this progress was thanks to Wernher von Braun's Peenemünde team, who tackled problems no one had cracked before. They built engines packing over twenty-five tons of thrust, powerful enough to carry a warhead miles above the Earth. Their guidance systems, using gyroscopes and accelerometers, kept the V-2 on course even when wind pushed against it, something earlier rockets struggled with.

Their true achievement was combining lightweight design, powerful engines, and aerodynamic features, enabling the V-2 to reach speeds of 3,500 miles per hour and heights over fifty miles. This made it the world's first true ballistic missile, the direct ancestor of rockets that would eventually take humans to the Moon.

However, this technological breakthrough came with a heavy human cost. V-2 rockets were produced in harsh underground factories using forced labor, and sadly, more people died building them than in their use as weapons.

When the war ended, the race shifted from territory to the brains and blueprints behind these rockets. Through Operation Paperclip, the United States brought over 1,600 German rocket scientists, including von Braun's team, to New Mexico, where they test-launched the V-2 once again. Soon, the team moved to Huntsville, Alabama, transforming their military skills into the powerful Redstone and Saturn rockets of the space age.

Meanwhile, the Soviet Union was also chasing the same rocket secrets. They set up test ranges at Kapustin Yar and Baikonur and reverse-engineered the V-2, turning German technology from a weapon of war into a key part of their space ambitions—a rivalry that would change history.

Today, rockets serve many purposes. They carry warheads and satellites, monitor rivals from above, and send political messages with every signal beamed from orbit. During the Cold War, the sky became a new frontier of competition, with technology becoming a key measure of a nation's strength.

A Beep Heard Around the World

Everything changed on October 4, 1957. An 83.6-kilogram metal sphere, Sputnik 1, lifted off the Kazakh steppe atop an R-7 rocket and broadcast a steady "beep-beep" to the world. Humanity now had its first artificial moon, circling Earth every ninety-six minutes for twenty-one days. Those beeps signaled more than Soviet success; they marked the beginning of the Space Age.

Sputnik 1 proved that rockets, orbital mechanics, and guidance systems worked in practice. Launching objects into orbit had become possible, and concepts such as escape velocity and orbital insertion became urgent topics rather than theoretical ones.

The Soviets quickly followed with Sputnik 2, this time carrying the Moscow street dog Laika, the first living creature to orbit Earth. Laika's journey ended in heartbreak when she died from overheating and stress, but it showed us that life could survive the violence of launch, and the initial conditions of weightlessness, helping open the door to human spaceflight.

For the United States, Sputnik was a wake-up call. The thought of Soviet satellites passing overhead raised difficult questions, and the imagery of objects orbiting above American cities led to a swift boost in spending, civil defense, and scientific focus.

Here Comes NASA

America responded decisively by creating the National Aeronautics and Space Administration (NASA) in July 1958, unifying space exploration under civilian leadership. The agency embodied America's commitment to space leadership, and programs like Explorer, Mercury, and Gemini soon followed.

Early days weren't perfect. The Vanguard TV3 rocket fizzled to the pad, lampooned as "Kaputnik," "Flopnik," and "Stayputnik." But resilience paid off: Explorer 1 reached orbit in January 1958 aboard a Juno I rocket designed by von Braun's team, and made an accidental but monumental discovery—the Van Allen radiation belts.

These belts are invisible donuts of radiation surrounding Earth where high-energy particles from the Sun get trapped by the Earth's magnetic field. Explorer 1 found these because its Geiger counter quit working at high altitude—overwhelmed by radiation.

The belts are named for physicist James Van Allen (1914–2006), whose team's instruments had been overwhelmed. His work helped establish satellites as scientific tools, driving the development of planetary probes and space-weather monitoring.

Science often rewards the bold and, occasionally, the fortunate.

The First Humans in Space

On April 12, 1961, Yuri Gagarin circled Earth in under two hours in his Vostok 1 capsule. With a shout of "Poyekhali!" ("Let's go!"), he became a symbol for a generation, showing humans could survive the stresses of launch and the experience of microgravity.

He returned to Earth by ejecting from his capsule at roughly seven kilometers altitude and landing by parachute.

The United States answered quickly. Alan Shepard (1923–1998), piloting Freedom 7, became the first American in space just weeks later. His fifteen-minute flight was a suborbital hop, shorter than Gagarin's—but piloted all the way and safely splashed down, setting new standards for crew control and recovery.

These flights highlighted sharp contrasts: Soviet approach emphasized risk-taking and rapid achievement; the American approach emphasized pilot authority and controlled testing. Each approach came with victories, both igniting a race that would define a generation.

The Race Accelerates

Things really started to speed up from here. Gherman Titov (1935–2000) became the first person to take a nap (and get sick) in orbit. John Glenn orbited the globe and proved humans could work high above Earth. Valentina Tereshkova became the first woman in space, and Alexei Leonov (1934–2019) performed the world's first spacewalk, nearly becoming trapped outside when his suit ballooned in the vacuum. He bled air from the suit to regain flexibility, accepting the risk of decompression sickness as his core temperature rose from the strain.

These steps laid the groundwork for lengthier missions and more ambitious targets: the Moon was now in sight.

Reaching for the Moon

While human spaceflight captured worldwide interest, the Soviets also made incredible strides with the Luna program, a fleet of robotic explorers working hard to reach important lunar milestones. Luna 1 didn't quite reach the Moon but made history as the first spacecraft to escape Earth's gravity. Luna 2 successfully landed on the Moon—a historic first. Luna 3 took the very first photos of the hidden far side of the Moon.

Throughout the 1960s, the Luna missions kept advancing quickly. Luna 9 achieved the first soft landing, and Luna 10 became the first artificial lunar satellite. Later missions brought back soil samples, mapped gravitational differences, ran experiments, and showed that robots could explore places humans weren't yet able to go.

While the achievements of Luna sometimes get less recognition compared to the human spaceflight missions, they played a vital role in making real science happen—and provided engineers critical knowledge in the race to reach the Moon.

Chapter 14

Apollo Era

Presidential Challenge

By the late 1960s, humanity finally did the impossible—we walked on the Moon. For the first time, boot prints marked the dust of another world, turning moon gods and myths into history and putting our machines, and our hopes, on another world.

Just weeks after Yuri Gagarin orbited Earth, President Kennedy challenged America: "I believe that this nation should commit itself to achieving the goal, before this decade is out, of landing a man on the Moon and returning him safely to the Earth." At Rice University, he doubled down—the words still echo: "We choose to go to the Moon in this decade and do the other things, not because they are easy, but because they are hard." That speech lit the fuse on one of humanity's greatest technological adventures.

Saturn V: The Workhorse

Saturn V was more than a rocket—it was a game-changer. Built by Wernher von Braun at NASA's Marshall Space Flight Center, it towered 363 feet, making it taller than the Statue of Liberty. Until SpaceX's Starship, nothing had more power or reliability.

This was the first rocket designed for lunar travel, not warfare. It could hurl 140,000 pounds to low Earth orbit—setting a record that stood for nearly fifty years. Saturn V's innovations became rocketry's gold standard, shaping every heavy-lift rocket that followed.

How powerful was it? Unbelievably immense. Five roaring F-1 engines could devour three tons of fuel every second. In under a minute, they would drain an Olympic swimming pool. Combustion instability– violent pressure waves that could tear engines apart– nearly ended the program. But NASA engineers tamed those destructive shock waves by adding baffles to calm the boiling inferno inside every engine.

Saturn V's newly developed J-2 upper-stage engines could be restarted in space–which would be essential for lunar missions. Across thirteen epic launches, with three million parts and 1960s computers (!), Saturn V was extraordinarily reliable as it never failed to deliver its crews and cargo.

No mission tested Apollo's engineering like Skylab's launch in 1973, when its micrometeoroid shield tore away during liftoff vibrations, pinning a solar panel. Without proper shielding and power, Skylab faced overheating and power shortage. In an audacious fix, a swift rescue crew consisting of Pete Conrad (1930-1999), Joseph Kerwin, and Paul Weitz (1932-2017) zoomed up in a Saturn IB, deployed a sunshade, and freed the jammed panel, saving America's first space station.

The Apollo Missions in Sequence

The journey to the Moon wasn't a straight path. The early Apollo missions built up the experience needed for the Moon landing.

Apollo 7 launched in October 1968 as the first crewed mission after the Apollo 1 tragedy. During this eleven-day Earth orbit, the crew tested the redesigned Command Module and broadcast the first live television from an American spacecraft, though crew illnesses created tensions with ground control.

In December 1968, Apollo 8 made history as the first crewed mission to leave Earth's orbit. Commander Frank Borman (1928-2023), Jim Lovell (1928-2025), and William (Bill) Anders (1933-2024) traveled to the Moon and orbited it ten times, capturing the iconic "Earthrise" photograph. Their Christmas Eve Genesis reading from lunar orbit resonated worldwide, concluding with Borman's words: "And from the crew of Apollo 8, we close with good night, good luck, a Merry Christmas–and God bless all of you, all of you on the good Earth."

Apollo 9 took a significant step by testing the Lunar Module in Earth orbit in March 1969. James McDivitt (1929-2022) and Rusty Schweickart conducted the first test of the complete Apollo spacecraft configuration, including docking maneuvers and a spacewalk to test the lunar spacesuit and backpack.

Apollo 10 served as the final dress rehearsal in May 1969. Thomas Stafford (1930-2024), John Young (1930-2018), and Eugene Cernan (1934-2017) brought the Lunar Module within 15.6 kilometers of the lunar surface, testing all procedures except landing. They overcame a potentially catastrophic Lunar Module gyration, but it confirmed readiness for Apollo 11.

The Landing: Apollo 11

Apollo 11 launched on July 16, 1969, carrying Neil Armstrong (1930-2012), Buzz Aldrin, and Michael Collins (1930-2021) on humanity's greatest journey–600 million people watched worldwide. After entering lunar orbit on July 19, the Lunar Module Eagle separated from the Command Module Columbia on July 20.

During descent, Eagle's computer triggered "1201" and "1202" alarms, signaling processor overload, NASA-speak for "the computer's freaking out"–but Mission Control cleared them to continue. With fuel running low, Armstrong manually guided Eagle past a boulder field. At 4:17 PM EDT, he announced: "Houston, Tranquility Base here. The Eagle has landed."

At 10:56 PM EDT, Armstrong stepped onto the lunar surface with his famous words: "That's one small step for man, one giant leap for mankind." Over twenty-one hours, Armstrong and Aldrin collected 47.5 pounds of moon rocks, deployed scientific experiments, took photographs, and planted the American flag. They left a plaque that read "Here men from the planet Earth first set foot upon the Moon. July 1969 A.D. We came in peace for all mankind." After returning to Columbia, they splashed down safely in the Pacific Ocean on July 24.

> **Shenkin's Gauge**: Some moments are just too good to make up. In the late 1960s, Gerald Shenkin ran the show at Simmonds Precision in Long Island City, Queens–one

of those unsung companies that built components of the Apollo program behind the scenes. Simmonds supplied a critical fuel measurement device for the lunar module, per NASA specifications, engineered to warn Neil Armstrong the moment fuel got dangerously low—a warning he'd really need during that nail-biting Apollo 11 landing.

But after the safe touchdown and euphoria died down, NASA discovered the warning never sounded. Not once. Panic, meetings, and a bunch of finger-pointing. Then, a few months later, Shenkin got a heated call from someone at NASA, probably an engineer still losing sleep over it. "We need that gauge back for analysis—now."

Without missing a beat, Gerald replied in his signature deadpan style "Sure. Just one problem—it's on the Moon." Silence. Then a sheepish, "Oh... yeah." Click. Nobody ever called about that gauge again.[1]

Beyond Apollo 11

Apollo 12 blasted off on November 14, 1969, with one clear goal: prove that humans could land exactly where they aimed. Just thirty-six seconds after liftoff, lightning struck the Saturn V not once, but twice, causing multiple system failures. Flight controller John Aaron saved the day with his legendary call, "Try SCE to Aux," and the crew kept flying. On November 19, Pete Conrad and Alan Bean stuck the landing—parking Lunar Module "Intrepid" just 535 feet from Surveyor 3, a robotic visitor left on the Moon since 1967. The astronauts hit their target, showed off pinpoint navigation, set up science experiments, grabbed 75 pounds of samples, and even brought back pieces of Surveyor 3 for analysis.

Apollo 13 is the stuff of legend (and a Hollywood blockbuster). In April 1970, an oxygen tank blew out 200,000 miles from home. Commander Jim Lovell, Jack Swigert, and Fred Haise Jr. had to scrap the landing and use their Lunar Module as a lifeboat. They improvised with everything they had—including jerry-rigged CO_2 scrubbers—pulling off a miracle return that the world would call a "successful failure."

---

[1] Story as recounted by Todd Shenkin, son of Gerald Shenkin, who was Plant Manager at Simmonds Precision in Long Island City, Queens, during the late 1960s.

Apollo 14, led by Alan Shepard in January 1971, was determined to finish what Apollo 13 started. Despite tricky docking maneuvers, Shepard (the oldest moonwalker, at age 47) capped the mission with a swing at lunar golf, sending two balls sailing across the gray landscape.

Then came Apollo 15, a mission all about exploration. In July 1971, Commander David Scott and James Irwin rolled out the lunar rover—covering seventeen miles, sampling the Moon's surface, and finding the famous "Genesis Rock." Apollo 16, followed in April 1972, traversing the lunar highlands. Astronaut Charlie Duke left a family photo on the Moon—a snapshot with cosmic staying power.

The grand finale: Apollo 17, in December 1972. Harrison Schmitt, the program's only scientist-astronaut, joined Commander Gene Cernan for three packed days of geology, breaking records for moonwalk time and sample collection. Before climbing the ladder home, Cernan scribbled his daughter's initials in the dust—the last human mark left on the lunar surface.

Taken together, the Apollo missions proved that with teamwork, innovation, and sheer determination, humans were capable of conquering the impossible: turning the Moon from a distant object into a world we could touch, explore, and—just maybe—someday call home.

What Apollo Gave Back to Earth

Apollo's technological legacy still shapes our world today. The Apollo Guidance Computer (AGC) compressed room-sized machines into something that fit in a briefcase, and powered in-flight calculations are now found in every GPS, phone, or car everywhere you go.

Mission engineers solved problems that are still relevant today: bulletproof reliability, user-friendly controls, and enabling multitasking—thanks to NASA's drive for ever-smaller microchips. Although each microchip initially cost $1,000 (about $8,890 now), NASA's large orders helped reduce costs while also raising the bar for reliability, playing a key role in launching the computer revolution.

How do you drill into moon rocks at -250°F, without an extension cord? You invent the Apollo Lunar Surface Drill (ALSD), a battery-powered cordless drill (using a Black & Decker's power head). And the Apollo space suits? They were created by Playtex's parent company, using the talented Playtex seamstresses to sew advanced fireproof, airtight fabrics that firefighters and medical tech companies now use.

Apollo's communication systems proved that voice, data, and live TV could be reliably sent between the Moon and Earth using redundant space-to-ground links, and the expertise gained fed into later satellite-based communications that underpin today's global broadcasting, navigation, and networking technologies.

But more than just machines and milestones, Apollo showed us what's possible when teams from government, academia, and industry come together with a shared purpose; their teamwork continues to inspire advancements in disaster response, vaccine distribution, and climate solutions here on Earth.

The message from Apollo still resonates strongly today: our home planet is just the beginning, and with teamwork and innovation, we can face and overcome any challenge—whether in space or right here at home.

Chapter 15

Winged Delivery

Orbital Workhorse

After the Apollo Moon missions ended in 1972, NASA set its sight closer to home: building a reusable spacecraft that could make spaceflight routine. The result was the Space Shuttle–three vehicles in one: a rocket during launch, a spacecraft in orbit, and an airplane on return. The first Shuttle–Columbia–proved this on April 12, 1981, when it successfully landed back on Earth.

The Space Shuttle quickly established itself as the go-to vehicle for building and fixing things in space. For the International Space Station (ISS), it served as both a contractor and a cargo transporter, carrying modules up above while astronauts skillfully operated its robotic arm like a cosmic crane. Through spacewalks and precise maneuvers, crews worked together to piece together the largest human-made structure in space–assembling modules, solar arrays, and life support systems hundreds of miles above Earth.

For Hubble, the Shuttle became an essential orbital workshop for repairs. When early images revealed a major flaw in Hubble's primary mirror, the Shuttle enabled what once seemed impossible–repairing a telescope in space. In over five missions, astronauts carried out complex repairs and upgrades, replacing old instruments with state-of-the-art technology. These missions turned Hubble from a potential failure into a phenomenal success, leading to discoveries that have expanded our understanding of the universe–finding the age of the cosmos, exploring distant galaxies, and uncovering new exoplanets.

This all highlights the Shuttle's incredible versatility as both a spacecraft and an orbital workshop. It demonstrated that complex construction and maintenance could be safely carried out hundreds of miles above Earth. And this has expanded our horizons for human spaceflight and scientific research, paving the way for future space stations, telescopes, and exploration missions.

NASA once had a dream of launching the Shuttle fifty times a year. While the reality was more modest—135 missions over thirty years, averaging about four to five launches each year—these missions still brought groundbreaking changes to what humanity could achieve in space.

Technological Legacy

The Shuttle's innovations didn't just stay in space—they came home with us. From keeping track of heartbeats in orbit to monitoring health on your wrist, the program's medical advances are now a part of our everyday lives through modern telemedicine and wearable devices that keep an eye on everything from heart rate to sleep patterns.

The breakthroughs in materials science were just as impactful, especially when it comes to safety and protection. We will talk more about this in the Space-Age Materials section later.

Perhaps the most significant contributions were in robotics. The Canadarm—our Shuttle's robotic arm—did more than just grab satellites; it opened the door to the future of surgery. Its incredible progress in remote control and automation led to tiny, precise robots that now perform surgeries inside the human body with amazing accuracy. These robotics innovations also helped manufacturing, bringing in systems for detailed assembly and quality checks.

Plus, the Shuttle turned into a floating observer for Earth. Its special vantage point and advanced sensors gathered incredible data about our planet's climate, atmosphere, and weather. This information was important in tracking ozone levels, understanding climate change, and making weather predictions more accurate. The methods we developed for these missions continue to shape how we study and protect our environment today.

International Cooperation

The Space Race was all about outdoing rivals, but the Shuttle era marked a major change—building and working together. This program changed how countries interact in space, moving from Cold War rivalry to incredible global teamwork. Former competitors became friends, with NASA regularly welcoming astronauts from European countries, Japan, and even Russia after the Cold War ended. These diverse specialists brought a wealth of knowledge and new ideas to every mission.

It really fostered worldwide technical teamwork. NASA partnered with the European Space Agency (ESA) on Spacelab missions, and the Canadian Space Agency contributed the amazing Canadarm. These weren't just simple arrangements—they involved joint planning, shared research goals, and teamwork in developing new technologies. This international collaboration created useful protocols and relationships that later helped with complex projects like the International Space Station and modern commercial space efforts.

The teamwork model from the Shuttle era changed how countries explore space, setting a great example for today's international projects, whether exploring Mars or building the lunar gateway. Moving from competition to cooperation has made space exploration smarter, more innovative, and a benefit for everyone.

Learning from Tragedy

Despite its many achievements, the Shuttle always carried some risks with every launch. The Space Shuttle program faced two devastating accidents that forever changed space exploration. One of these was the Challenger disaster on January 28, 1986, just seventy-three seconds after lift-off. That morning was unusually cold, and a rubber seal called an O-ring on the right rocket booster became too stiff to do its job. This allowed hot gases to escape, burning through the booster's metal and causing the huge fuel tank to explode. The shuttle broke apart in midair, tragically taking all seven astronauts, including teacher Christa McAuliffe, with it. Engineers had warned that the O-rings might fail in cold weather, but the launch went ahead anyway. This heartbreaking event led to a complete overhaul of NASA's safety procedures.

The Columbia disaster on February 1, 2003, was a tragic event during re-entry. A piece of foam insulation had come loose during launch and harmed the shuttle's thermal protection system, which caused the spacecraft to break apart upon re-entry. Sadly, all seven crew members lost their lives in this heartbreaking accident.

These tragedies prompted meaningful changes in NASA's safety practices. The agency now carries out more thorough pre-launch inspections, fosters better communication between engineers and management, and has established an independent safety oversight office. Most importantly, NASA nurtured a culture where every employee feels empowered to stop operations if they notice a safety concern. These important reforms are now widely adopted across the aerospace industry, reminding everyone that safety should always come before schedules.

Enduring Impact

When Atlantis completed the final mission (STS-135) in 2011, it marked the end of an era that spanned thirty years and 135 missions. The Space Shuttle's legacy continues to influence modern spaceflight in profound ways. Today's spacecraft like SpaceX's Crew Dragon and Boeing's Starliner incorporate many lessons learned from the Shuttle program, from thermal protection systems to launch abort capabilities.

The Space Shuttle era pioneered techniques in space construction that were essential for building the International Space Station, developed orbital repair and maintenance procedures that are still used today, and established protocols for long-duration spaceflight that inform current missions.

Most importantly, the Space Shuttle program changed our relationship with space. It demonstrated that space wasn't just a distant frontier to be reached, but a place where humans could work regularly, conduct groundbreaking research, and build permanent infrastructure. This shift in perspective paved the way for today's commercial space industry and our ambitious plans for lunar bases and Mars exploration. The Space Shuttle's greatest achievement might be reminding us that space isn't just a place to visit once, but a whole environment where we can build a lasting presence.

Chapter 16

Private Rocketeers

The Commercial Challengers

For much of the twentieth century, space was primarily the realm of governments: huge projects funded by public money, driven by national pride and Cold War tensions.

From Apollo to the Space Shuttle, exploring beyond Earth was a government-led effort—closely managed, seldom open to others, and incredibly costly. However, in the early 2000s, things started to change. A fresh wave of private companies stepped in, daring to challenge conventions and old engineering norms.

The Reusability Revolution

Leading that charge was SpaceX. Founded in 2002 by entrepreneur Elon Musk, SpaceX detonated the aerospace industry and turned everything on its head. With a mission to make humanity multiplanetary, the company took aim at the most sacred assumptions of rocket science: that launches had to be disposable, that space access had to be slow and bureaucratic, and that innovation had to creep forward at the pace of government procurement. SpaceX rewrote the rules—and then burned the old playbook.

The first breakthrough came with Falcon 1. After a string of public failures, SpaceX succeeded in 2008 in launching the first privately developed liquid-fueled rocket to reach orbit. It was a victory built not on deep legacy, but on relentless iteration. From there came the Falcon 9, which launched in 2010 with nine Merlin engines and a two-stage design tailored for payload flexibility. It was modular, scalable, and built to be reused—a radical departure from the single-use models that had defined the space age until then. Reuse would become the company's defining legacy.

In December 2015, SpaceX landed a Falcon 9 booster upright on solid ground. A year later, they did it again—this time on a floating drone ship in the ocean. The sight of a fourteen-story rocket descending, controlled and precise, was once science fiction. Now, it was operational. By the mid-2020s, Falcon 9 boosters had been reused over twenty times each. Launch costs plummeted. Reliability soared. Small satellite startups, universities, and even emerging nations now had access to orbit.

With Falcon Heavy in 2018, SpaceX delivered the most powerful rocket in the world. Capable of lifting 63,000 kilograms to low Earth orbit, it tripled the performance of its closest rivals. And for its debut? It didn't launch a weather satellite—it launched a red Tesla Roadster into interplanetary space, just to prove it could.

But even as Falcon 9 and Heavy took over commercial and government launches—including NASA resupply missions to the ISS—SpaceX was already building what came next: Starship.

Starship is a fully reusable, two-stage super-heavy launch system made of stainless steel and powered by methane-burning Raptor engines. Designed to carry over one hundred metric tons and one hundred humans—it's built for destinations far beyond Earth. The Moon. Mars. Maybe beyond. It's part rocket, part freight train, part colony ship. As of the mid-2020s, Starship is still in rapid testing at SpaceX's Starbase in Texas, with full-stack launches and flight trials stacking up alongside global anticipation.

In 2020, SpaceX became the first private company to launch humans into orbit with Crew Dragon Demo-2, restoring US crewed launch capability for the first time since the Shuttle. Since then, they've launched multiple missions under NASA's Commercial Crew Program, as well as Inspiration4—the first all-civilian orbital flight—and Axiom missions that pioneer private astronaut flights and commercial space station infrastructure.

A Constellation of Innovators

As SpaceX blazed its trail skyward, others took note. A constellation of innovators followed—small, agile companies determined to rewrite the rules of rocketry in their own way.

Virgin Galactic emerged as one of the most visible commercial routes to human spaceflight by focusing on suborbital journeys to the edge of space. Its approach drew directly from the legacy of SpaceShipOne, the experimental craft that won the Ansari XPRIZE in 2004 by reaching the one hundred-kilometer Kármán line twice within two weeks. Virgin Galactic licensed that architecture and began translating the prize-winning prototype into a reliable, passenger-ready system designed for frequent flights rather than one-off demonstrations.

SpaceShipTwo, flown as VSS Unity, carried that lineage forward while operating in a different regime. Instead of aiming for one hundred kilometers, Unity targets altitudes above fifty miles—the threshold that the United States recognizes as the beginning of space. Released from the twin-boom mothership VMS Eve high above New Mexico, Unity ignites its hybrid rocket motor, accelerates through Mach 3, and climbs to roughly 82 to 90 kilometers. At apogee, passengers experience several minutes of microgravity and sweeping views of Earth's curvature before the craft reconfigures its tail into the "feathered" aerodynamic mode and glides back to a runway landing. The sequence blends test-pilot experience with a commercial ambition: to make access to suborbital space feel closer to high-performance aviation than to traditional rocketry.

Scottish-born chief pilot David Mackay played a central role in proving the system. He became the first native-born Scot to fly in space during the 2019 VF-01 flight and later commanded Unity 22 in 2021, carrying Richard Branson and a small research team on a mission broadcast worldwide. Those flights marked the moment when Virgin's vision turned into a real, accessible, (and purchasable!) experience. Looking ahead, the company is developing its Delta-class spaceplanes—production-model vehicles designed for higher flight cadence and six paying passengers per trip—with first commercial service expected later in the decade.

Incidentally, Virgin and David were operating in a regulatory system that deliberately made room for them to experiment. The rules around them were written on the assumption that private teams would be the ones testing the edge of what was possible, and the government's job was to keep people safe without shutting those experiments down. Indeed, when David's team strayed beyond their assigned New Mexico airspace, they expected to get into a lot of trouble. Instead, the first response was a question: What would help you avoid this next time? After they explained to the FAA what went wrong, why it went wrong, and what they were going to do to prevent it from happening again in the future, they were given more airspace, not less.

While Virgin explored the exciting frontiers of human spaceflight, others turned their attention to a different challenge: the increasing demand for small satellites and the lack of rockets specifically designed for them. Rocket Lab took a fresh approach compared to traditional heavy-lift spacecraft, making space launches more accessible and customized to the needs of small satellite operators. Originally founded in New Zealand and now based in the United States, the company recognized early on that in the fast-growing world of microsatellites, speed was more important than size. Small spacecraft could be designed and built more quickly than the time it took to secure slots on larger rockets, so Rocket Lab developed Electron to meet that need directly.

Electron represents the idea that getting to space should be as reliable and straightforward as sending a package via a trusted courier, rather than relying on infrequent large rockets. Instead of competing with big launches, Rocket Lab focused on consistent scheduling, dedicated missions, and precise manufacturing. This approach led to the first commercially successful small launch vehicle designed for quick turnarounds. Universities, startups, and sensor teams could now put their satellites into orbit on their own timetable instead of waiting for months or even years for a shared launch that might not perfectly suit their mission. Electron's success proved that space access can be a dependable service with predictable timing.

As the satellite industry evolved, Rocket Lab recognized that future growth would come from deploying entire constellations, not just individual satellites. Large satellite operators increasingly need dedicated launches for multiple spacecraft, with quick replacements and tight control over orbital paths. To meet this need, they developed Neutron, a medium-lift, reusable rocket that bridges the gap between small, dedicated launches and larger vehicles like Falcon 9.

Neutron embodies Rocket Lab's belief that future space logistics will live in the middle ground between tiny, dedicated launches and giant megaconstellations. It is built to carry clusters of satellites, national security payloads, and heavier commercial spacecraft, while keeping the lean operations that made Electron so successful. Its fully reusable first stage and tapered carbon-fiber structure are paired with an unusual "Hungry Hippo" fairing: instead of throwing the nose cone away, Neutron's fairing stays attached, opening and closing like a set of jaws around the payload so it can be reused. All of these choices aim at quick turnarounds and minimal ground infrastructure–bringing Electron's agility into a much more powerful class of rocket.

Strategically, Neutron positions Rocket Lab to serve the middle market: customers who need more capacity than Electron offers but don't require the size or expense of the largest rockets. It builds on Rocket Lab's core idea that timing control is just as important as raw power–adding flexibility, cadence, and reusability to how constellations are built and kept operational.

While Electron empowered small teams to launch satellites on their own terms, Neutron expands that freedom to companies developing future orbital infrastructure. In this way, Rocket Lab is moving from just being a courier to becoming a creator of orbital logistics infrastructure, shaping a launch ecosystem where quick response and reusability are central to strategy.

Then came Relativity Space, with a different take altogether. Rather than iterating on existing rocket-manufacturing methods, Relativity bet on something radical: building entire rockets using AI-guided 3D printers. Its Terran series of rockets–printed from metal alloys and designed to reduce parts count from 100,000 to less than 1,000–promised faster manufacturing, fewer failure points, and an eventual path toward autonomous off-world construction.

Relativity's hypothesis was that if you can print an entire launch vehicle on Earth, you can print structures on Mars, the Moon, or any other world where raw materials and energy exist. Manufacturing became not merely a process but a gateway to off-world industry.

Firefly Aerospace joined the race with a focus on medium-lift rockets and lunar payload delivery. Rising from the ashes of a previous company and now backed by private equity and government contracts, Firefly's Alpha rocket added capacity where neither SpaceX nor Rocket Lab played directly–delivering science, sensors, and landers for lunar and orbital missions. Firefly Aerospace recently achieved a historic milestone with its Blue Ghost Mission 1, which successfully soft-landed a robotic lunar lander on the Moon on March 2, 2025. This was the first fully successful commercial soft landing on the Moon. The lander delivered 10 NASA scientific and technological payloads to the lunar surface in the Mare Crisium basin, a large lunar impact basin.

The response from legacy aerospace was immediate and overdue. Boeing scrambled to keep the Starliner capsule on schedule. Northrop Grumman doubled down on Cygnus cargo missions and solid-fuel boosters. United Launch Alliance, once dominant, found itself adapting to an entirely new tempo–racing to finish its Vulcan Centaur rocket while watching SpaceX land and relaunch the same Falcon 9 booster multiple times.

Entire nations also felt the shift.

China accelerated its ambitions with the Long March series, its Tiangong space station, and multiple lunar sample return missions. The Indian Space Research Organization (ISRO) stunned the world with cost-effective Mars and lunar missions, proving that space success no longer required billion-dollar budgets. Their Vikram lander and Chandrayaan probes became national pride projects with global implications.

Through all of this, one pattern emerged: space was no longer slow. The cadence of exploration had changed. In the 1960s, a single launch could dominate headlines for weeks. Today, rocket launches are streamed on YouTube, followed on X (formerly Twitter), and scheduled alongside sports broadcasts. It's not hyperbole to say they happen weekly. Sometimes, daily. As of the mid-2020s, SpaceX alone launches more orbital missions per year than any national space program—including China, Russia, and even the US government.

What was once an arms race has become an uptime race.

The New Race

It's a new kind of space race—more about frequent flights and commercial cadence and less about flags or Cold War symbolism.

It doesn't stop in orbit. Starlink is changing how we access the internet, especially in rural and underserved areas. Thousands of satellites launched on Falcon 9 are now forming a global communication network built entirely in-house.

What started as a Silicon Valley gamble has grown into the heart of a thriving space economy. Space is no longer just for nations; it's becoming a vibrant space for businesses, tourists, universities, and dreamers alike. From space stations to lunar logistics, and from Mars habitats to orbital manufacturing, the next chapter of space exploration is driven by missions and innovation, not just flags. The launchpad itself is becoming like a startup hub with a flame trench.

In 1944, rockets thundered over cities. By 1969, they carried astronauts into space. Today, in the 2020s, rockets bring internet, cargo, civilians, and visionaries. From the early fire arrows to Falcon 9, the story of rocketry is one of both national and commercial effort. The New Space Age is already here, visible in the hoodie-wearing explorers walking out of a hangar in South Texas, looking up.

Chapter 17

Anatomy of a Liquid Rocket

Controlled Explosions of Fury

In Science Class, we covered some of the fundamentals of thermodynamics, fluid mechanics, and aerodynamics used in rockets and spaceflight. Now is the time to see how they come together in perhaps the ultimate application imaginable: the rocket engine.

This amazing feat of engineering creates a controlled explosion of absolute fury where thermodynamics manifests in temperatures hot enough to melt steel, where fluid mechanics governs the flow of propellants, and where aerodynamics shapes the supersonic exhaust plume. All while not destroying itself.

Most rockets these days are generally liquid-fueled engines so we will use that as a starting point to explain how rockets work. From propellant tank to nozzle, from injector plate to exhaust plume, we will see how this is a very balanced process of heat and force and how engineers harness these fundamental forces to create one of humanity's most powerful machines.

Welcome to the fun part of rocket science. Let's light this thing up.

A Rocket's Heart

At the heart of every rocket's journey from the launch pad to space is the rocket engine cycle—a complex, finely tuned process that transforms stored energy into the powerful thrust that propels a rocket skyward. This cycle begins with the storage and management of cryogenic propellants in advanced tanks (think giant thermos flasks), which keep fuel and oxidizer super-cold and ready for the moment of ignition. Before launch, engineers prepare the entire system, gradually chilling and pressurizing the tank structures so they can withstand extreme temperatures and enormous mechanical stresses.

Once ignition begins, the engine's turbopumps act like high-powered blenders, forcing the fuel and oxidizer into the combustion chamber at incredible pressures and speeds. In this chamber, the propellants mix and burn, rapidly transforming into a searing hot, high-pressure gas. This gas is then directed through finely engineered injector plates and accelerated out of a nozzle, converting heat and pressure into supersonic exhaust for maximum thrust.

Throughout the flight, a network of systems maintains optimal conditions. Pressure is regulated, ullage gases keep tank walls stable, and advanced baffles ensure liquid propellants always flow uninterrupted to the engine, even in zero gravity. The rocket's steering controls–gimbaled engines, aerodynamic fins, and reaction thrusters–work together to guide the vehicle along its calculated trajectory, adjusting to ever-changing altitude, speed, and environmental conditions.

The entire cycle is a testament to modern engineering: each launch coordinates temperature, pressure, mechanical precision, and powerful chemistry, all designed to tame explosive forces into controlled, reliable motion. This introduction sets the stage for a deeper look at the extraordinary technologies and innovations that make rocket propulsion possible.

Let's dig in a bit deeper into each of these parts of the rocket engine cycle.

Thermos Flasks

On a really cold day hiking in the mountains, it is really nice to have a wee cup of warm soup from a flask you have carried up the hill with you. On a scorching hot day in the middle of summer, you may be gasping for some ice-cold water (or juice, or your preferred beverage of choice) from your water bottle to cool you down. In both cases, you may be carrying a flask—an insulated double-walled container with a vacuum in between, and a reflective liner, to reduce heat loss (or gain). You would also be well advised to keep the lid tightly sealed, preventing further heat exchange, so that the drink's enthalpy (heat content) remains constant over time and to keep the drink the way you like it.

Rocket propellant tanks are just like giant thermos flasks—giant storage containers that hold the chemical lifeblood of the vehicle. Unlike on Earth, where oxygen is readily available, rockets must carry both fuel and oxidizer for combustion in space.

These cryogenic tanks are marvels of modern engineering, featuring multi-layered specialized insulation and aluminum-lithium alloys designed to endure the extreme conditions of spaceflight. These materials must maintain their structural integrity at temperatures approaching absolute zero and under the intense pressures and mechanical stresses of launch. The tank walls themselves are made to be as thin as possible to reduce weight while remaining strong enough to contain their volatile contents.

Managing these super-cold liquids is an even greater challenge than merely containing them. As the propellants warm during flight, they expand and vaporize, generating pressure that could compromise the tank's integrity if not properly controlled. Engineers have developed pressure regulation systems that continuously monitor and adjust conditions within the tanks.

To maintain optimal pressure as the tanks empty during flight, "ullage management" injects Helium or Nitrogen gas into the emptying tank space, ensuring proper pressurization and preventing the tank walls from buckling under the enormous external forces. This becomes even more critical in the zero-gravity environment of space, where the propellants can float freely within the tanks. To combat this, designers incorporate intricate networks of anti-slosh baffles and specialized screens throughout the tank structure that keep the propellants properly oriented toward the feed lines, so they flow consistently to the engines when needed.

Originally, anti-slosh baffles were developed in the 1940s and 1950s and have evolved since then. This has now been adopted in the fuel tanks of Formula 1 racing cars to prevent the fuel from sloshing around as the car hits high-speed corners and breaks into slower corners, which becomes even more of an issue as the fuel is burned during the race.

Chilling Out

Preparing for launch is a highly managed process called "chill down operations," where engineers gradually introduce small amounts of cryogenic propellant to cool the tank systems and associated plumbing. This approach helps avoid thermal shock, which could harm delicate components–similar to warming a frozen glass slowly with lukewarm water to prevent cracking–except in reverse and at much colder temperatures. If not controlled properly, an unexpected and dramatic phenomenon can occur–*geysering*.

Much like the famous Old Faithful geyser in Yellowstone, geysering occurs when trapped liquid heats unevenly, creating vapor bubbles that rise, burst, and send pressure surges through confined spaces. During rocket fueling, a similar process can happen when extremely cold liquids like liquid oxygen or hydrogen flow through warmer fuel lines into the tanks. The cold propellant near the walls warms up slightly and rises, while the colder liquid sinks, creating a convection cycle. As the warm layer grows, it boils, producing bubbles that suddenly burst upward through the lines–creating pressure spikes that can shake the entire fueling system. Just like Old Faithful's eruptions are spectacular yet natural, these propellant geysers require careful management to avoid dangerous pressure spikes and protect the rocket components.

Venting Before Violence

We have all seen the striking visual of a rocket sitting on its launch pad with billowing white clouds cascading down its sides. These clouds are not smoke or steam, as many assume. They are actually cryogenic vapors, primarily nitrogen and oxygen, being strategically vented from the propellant tanks.

This venting is important for maintaining the correct internal pressure and temperature conditions. Without this, the super-cold propellants could either build up too much pressure or warm up beyond acceptable levels, either of which could compromise the mission's safety. Also, boom!

So these giant thermos flasks have to continuously maintain a delicate balance of temperature, pressure, and stability from pre-launch all the way through the violence of lift-off until the tanks are empty in the emptiness of space.

Daily Cryogenic Uses

The development of these giant thermos tanks for rockets–designed to hold liquids so cold they can freeze steel–has been a remarkable triumph of space engineering and has also impacted many areas of everyday life. In building tanks that can safely store super-cold fuels like liquid oxygen or liquid hydrogen, rocket scientists faced enormous challenges: preventing heat from leaking in, controlling pressure, and maintaining extreme temperatures over long periods. The solutions they created have since found their way into medicine, food storage, industry, clean energy, and even global shipping.

One of the most important spinoffs has been in the area of medical storage. The insulation and temperature-control ideas originally used for rocket fuel tanks have evolved into the technology we now rely on for modern cryogenic freezers and transport containers. These are incredibly important for keeping delicate materials like biological samples, vaccines, and certain medicines safe. During the COVID-19 pandemic, the very technology that once took astronauts into space was working behind the scenes to keep mRNA vaccines stable at temperatures well below freezing.

The food industry has also felt the impact. By applying the same insulation techniques that keep liquid oxygen from boiling away in a rocket's tank, industrial freezers and cold storage warehouses can operate more efficiently than ever. This means that frozen food stays fresh longer, energy costs are reduced, and less food goes to waste before it reaches consumers.

In heavy industry, companies that store gases such as nitrogen, helium, or argon can take advantage of tank designs that were originally perfected for aerospace. Features like multi-layer insulation, pressure control systems, and specialized materials from rocket programs now enable large amounts of these gases to be stored and transported with minimal loss. This makes processes more dependable and cost-efficient.

Cryogenic tank innovations have even helped drive the clean energy revolution. Storing liquid hydrogen for rocket launches is quite similar to storing it for hydrogen-powered vehicles and fuel cell systems. Thanks to lessons learned in spaceflight, this challenging task is becoming more practical and is helping hydrogen emerge as a viable alternative to fossil fuels.

The shipping industry has also benefited a lot. Large carriers that transport liquefied natural gas (LNG) across oceans use cryogenic containment systems that are based on rocket technology. The same engineering that keeps propellants stable during rough explosions during launches now also helps keep LNG at the right temperature for long trips spanning thousands of miles.

In each of these cases, the effort to send rockets safely into space has surprisingly but positively influenced life here on Earth. It's a wonderful reminder that exploring space often leads to new innovations we enjoy daily, sometimes without even noticing their space origins.

Madman's Blender: The Art of Controlled Violence

At the heart of every rocket engine sits a component that works like the world's most extreme and dangerous blender: the turbopump. Without it, the rocket's fuel and oxidizer would just sit uselessly in their tanks, like a very expensive popsicle full of cryogenic soup.

The turbopump rips cryogenic liquids from their tanks and rams them into the chamber at pressures higher than the deepest ocean, and spinning faster than any household appliance could handle.

The basic structure is a turbine (similar to a small jet engine) that spins a shaft at a mind-boggling 90,000 times per minute, which then drives a pump that pushes fuel with enough pressure to crush a car and temperatures that can melt steel. The turbopumps are among the most stressed components in the entire rocket operating under incredible loads, temperatures, and rotational speeds, and must work flawlessly.

The impeller blades are crafted from resilient superalloys that can handle extreme heat and shattering forces. They're cut to be razor-thin yet remain strong and durable. Keep in mind, if a single blade develops a crack, or if there's a slight timing issue, it could tear the pump apart and take the entire vehicle with it.

When it comes to bearings, standard lubricants just won't do–they could freeze up in the cold or even overheat and ignite in the heat. Instead, alternatives like helium cooling systems keep the bearings at just the right temperature. While some experimental setups have explored magnetic levitation, where parts float without touching, most actual rocket turbopumps today rely on special fluids or gases to ensure everything spins smoothly and safely.

Every launch is a test of metallurgy, balance, and finely controlled violence. The evolution of turbopumps traces back to steam power, where basic mechanical pumps laid the foundation for moving liquids under pressure. The first major breakthrough for rockets came with the V-2 rocket program during World War II, where Werner von Braun's team created a hydrogen peroxide-powered turbopump generating 600 horsepower–modest by today's standards but very advanced back then.

The Space Race brought rapid advancement. The Apollo program's F-1 engine turbopump was particularly impressive, delivering 55,000 horsepower–more powerful than the engines of a Second World War destroyer. Soviet engineers pushed in a different direction, pioneering staged combustion engines such as the NK15 and RD170, which fed turbine exhaust back into the main chamber to squeeze more performance from each kilogram of propellant.

By the 1970s, the Space Shuttle's RS-25 engine marked another milestone: its fuel turbopump spun at 37,000 rpm while handling near-absolute-zero liquid hydrogen, with a separate system for liquid oxygen to keep the two propellants isolated until combustion.

Staged Combustion: The Heart of the Rocket

If the turbopump is what pushes the fuel through a rocket engine, then what powers the turbopump? A common and important answer is staged combustion. In this process, a small amount of propellant is pre-burned in a preburner, creating hot, high-pressure gases that spin the turbine in the turbopump. Instead of being expelled to the atmosphere, the gases flow into the main combustion chamber, where they burn alongside the remainder of the fuel delivered by the turbopumps. This setup is a very efficient use of the propellant to keep the engine running smoothly.

Over the years, there have been three main forms, each coming with its own set of trade-offs.

Back in the 1950s, the Soviet Union pioneered the use of oxygen-rich staged combustion in their rocket engines. Their preburners burned a small amount of fuel with a lot of liquid oxygen, creating a super-hot, oxidizing gas that powered the turbines and then fed into the main chamber. This approach achieved incredible efficiency, though it also created very harsh conditions inside the plumbing. To prevent the engines from damaging themselves, engineers had to develop special alloys, coatings, and cooling methods. Mastering this challenging chemistry allowed them to produce some of the most powerful and efficient kerosene engines ever created.

In the 1970s, the United States came up with fuel-rich staged combustion, which was famously used in the Space Shuttle's RS-25 engines. In this approach, the preburner runs with excess fuel and less oxygen, producing a hot gas that is gentler on metals and seals, reducing corrosion. While it was not as efficient as oxygen-rich designs, it was easier to develop. This successful approach inspired variants in engines like Japan's LE7 and India's CE7.5, showcasing its influence across the world.

The latest advancement in staged combustion is the full-flow system, which divides the process into two parts. It uses one preburner that's rich in oxygen and another that's fuel-rich, ensuring all propellant passes through turbines before reaching the main chamber. SpaceX's Raptor engine, which debuted in 2019, is the best-known example of this system where every kilogram of propellant does double the work–first helping power the turbines, then serving as fuel or oxidizer in the main burn. While the machinery is extremely complex, the benefits are very impressive: incredible efficiency, extremely high chamber pressure, and engines that can be reused multiple times.

Alternate Turbopumps

Rocket Lab changed small rocket design by creating a totally different kind of engine with their Rutherford engine. Instead of using the usual gas-powered turbopumps, it drives its pumps with electric motors powered by lithium-polymer batteries–similar to those in modern electric cars but made for space travel.

These electric pumps come with many perks. They are simpler, with fewer moving parts and no need for complex plumbing or pre-burners, making them more reliable and easier to produce. Plus, they produce less heat since there's no hot gas turbine, which helps reduce thermal stress on the engine. What's really cool is that electric pumps give engineers much more precise control over fuel flow–like turning a dimmer switch instead of flipping a light switch. This makes it easier to finely tune performance during flight and use fuel more efficiently, adjusting exactly when and how much propellant to deliver, even when conditions change.

Another benefit is that nearly all the propellant can, in principle, be used for thrusting, unlike traditional turbopumps, which burn some propellant simply to power the pumps. Think of it like comparing a hybrid car that uses some gas to generate electricity, to a fully electric car that can put all its stored energy into motion.

While this electric pump system isn't yet ready for bigger rockets like Falcon, it has made a big difference for smaller launch vehicles like Rocket Lab's Electron. It's a wonderful example of how rethinking basic rocket parts can lead to exciting new solutions for specific needs. Even as the technology matures and becomes more dependable, pumpfed engines remain among the most demanding parts of rocketry–a reminder of how much creativity it takes to turn frozen propellants into controlled fire and keep pushing the boundaries of spaceflight.

Turbopumps on Earth

The extreme engineering demanded by rocket turbopumps has spilled over into many parts of everyday life, helping to reshape industries from medical technology to power generation.

In medicine, miniature heart pumps and blood circulation devices borrow directly from turbopump principles. Designs first created to push cryogenic propellants at high pressure have been adapted into compact, reliable artificial heart pumps that help keep patients alive.

Power generation has also benefited from advanced materials and bearing designs originally developed for rocket turbopumps. Those lessons in handling extreme temperatures and pressures have improved the efficiency and durability of industrial gas turbines and other plant equipment.

Water systems gain from the same heritage. Modern treatment facilities and largescale irrigation networks use pump architectures and control techniques with roots in space program turbomachinery, improving the efficiency of water distribution and conservation.

Manufacturing methods refined for turbopumps have influenced the design of industrial mixing and processing equipment, contributing to food production systems that are both more efficient and more dependable.

Perhaps most far-reaching are the computer modelling and simulation tools created to design rocket turbopumps in the first place. These high-end fluid-dynamics software systems are now standard across engineering, helping to optimize everything from car engines to wind-turbine blades.

Hell's Showerhead

After cryogenic propellants are forced through turbopumps at extreme pressures, they converge at the injector plate, one of the most critical components for preventing catastrophic failure.

This disk contains hundreds of tiny holes that spray fuel and oxidizer in precisely designed patterns. If the mixing is wrong, propellant can pool before ignition and produce a "hard start," an explosion strong enough to destroy the engine right on the pad. Even in normal flight, poor injection can trigger "combustion instability events" (informally known as *boom time*), where pressure waves create violent resonances that can tear the engine apart in milliseconds.

The most common design is the showerhead injector, which really does work like an advanced shower head with hundreds of tiny holes. When fuel and oxidizer spray through these holes, they intersect to create a fine, efficiently combustible mist—much like two garden hoses spraying into each other.

Modern injector plates took shape in the 1940s with the German V-2 rocket program when engineers realized that successful combustion depended critically on propellant atomization and mixing before they burn. They found that droplet size, spray angle, and injection velocity all mattered: large droplets or uneven mixing could destabilize combustion and destroy the engine. V2 engineers came up with a double-wall injector that used concentric rings of ports, sending fuel through inner holes and oxidizer through outer ones. The overlapping sprays created mixing zones where propellants broke into fine droplets before ignition, improving stability and performance compared with earlier, single-element designs.

Another major advancement came with the pintle injector, developed for NASA's Apollo Lunar Module and later adopted by SpaceX for its Merlin engines. The design looks like one fire hose inside another and allows precise control of how fuel and oxidizer meet, which makes it easy to throttle and lends itself to reliable, reusable engines—which would be critical for both lunar landings and modern reusable rockets.

What really makes injector plates stand out is how well they endure their challenging environment. They handle liquids almost as cold as deep space (around −200 °C) and spray them into chambers that are hotter than a blast furnace (over 3,000°C). All of this occurs within a space roughly the size of a trash can lid, and it needs to operate perfectly within milliseconds.

The same ideas have reshaped car engines. Techniques for atomizing and mixing rocket propellants fed directly into the development of automotive fuel injection. As electronic control improved in the 1970s and 1980s, automakers moved from carburetors to injectors that spray precise patterns of fuel straight into the cylinders. Modern direct-injection systems are, in effect, a simplified descendant of rocket injectors: they meter fuel with high precision, guided by computers and built from durable materials. The result is that everyday cars are more powerful, more efficient, and cleaner than they would have been without lessons learned on the way to space.

Modern Innovation

Today's rocket companies are transforming injector plate technology by applying advanced thermodynamics, fluid mechanics, and cutting-edge tools such as Computational Fluid Dynamics (CFD), that can model how propellants' temperature gradients, pressure distributions, and turbulent mixing behaviors interact.

3D printing enables the creation of intricate geometries that optimize these thermodynamic and fluid properties—including mixing chambers and cooling channels impossible to manufacture through traditional methods.

Engineers model propellant atomization and mixing fluid mechanics in extraordinary detail before manufacturing. These simulations account for Reynolds Number, viscosity changes across temperature ranges, and complex turbulent flow patterns. This comprehensive understanding of fluid behavior helps predict and prevent dangerous combustion instabilities that could destroy engines.

Modern manufacturing has also enhanced our control over injector plates' temperature and cooling. Engineers now create more efficient cooling passages and better manage extreme temperature differentials between cryogenic propellants and combustion gases. This is very important because even minor deviations in propellant mixing or thermal management can cascade into catastrophic failures. The combination of advanced thermodynamics, fluid mechanics, and manufacturing has created injector plates that are exceptionally reliable.

The injector plate stands as the ultimate gatekeeper between controlled power and catastrophic failure in rocket engines—a masterpiece of engineering where success and disaster are separated by mere millimeters.

When properly designed and manufactured, it orchestrates a spray of atomized propellants producing smooth, reliable thrust. However, any slight imperfection transforms the engine into what engineers grimly call an "unscheduled rapid disassembly"—or impromptu steel confetti generator!

It really embodies rocket science at its finest: where humans have tamed the fury of rocket-grade hellfire into a predictable, manageable showerhead.

Caged Inferno

So now we get to the crazy part of the rocket where we take super-cold liquids (liquid hydrogen and oxygen, cold enough to freeze metal) and transform them into incredibly hot gas in a fraction of a second: the combustion chamber.

The combustion chamber isn't just "hot"—your kitchen oven is hot. This is apocalyptic. In mere milliseconds, it transforms frozen propellants into a plasma-spewing inferno at temperatures that make molten lava seem lukewarm. The interior reaches a staggering 3,300°C, which exceeds the surface temperature of Venus and as hot as some of the surface of some cooler stars (which can be as cool as 2,500°C!). For perspective, lava flows at about 1,200°C, while Venus's surface is only around 470°C. The Sun's surface, by comparison, is about 5,500°C. The atmosphere inside the chamber is not just hot—the pressure can be many times greater than what is found at the bottom of Earth's deepest ocean trenches.

During combustion, the gases that burst into existence are only partially ionized—their atoms lose some, but not all, electrons in the frenzy. These gases expand with tremendous force, creating a brilliant plasma that slams against the walls of the chamber. In that violent moment, the Second Law of Thermodynamics surges to life: chemical energy, locked away in ordinary molecules, is unleashed in a single, spectacular jailbreak. Atomic bonds shatter and reform at lightning speed, leaving behind an incandescent torrent of gas that races outwards far faster than anyone could utter the word "enthalpy."

But how does the chamber avoid melting? Engineers use a cooling system similar to how your body uses blood vessels to regulate temperature. They create tiny channels in the chamber walls where super-cold fuel flows through before being used, like a built-in cooling system. This process, called regenerative cooling, is the difference between a working rocket and an expensive explosion. They are using the fuel that is going to be burned to cool the fuel that is being burned!

The combustion chamber is built from extraordinary materials that can withstand conditions that would melt or destroy ordinary metals in seconds. Engineers use specialized metals based on their unique properties to create what is essentially a container for a miniature star: these materials must remain structurally sound while containing temperatures hotter than molten lava!

Three critical materials make this possible:

- Copper may seem ordinary, but it plays a very important role due to its exceptional thermal conductivity—it transfers heat extremely efficiently. Engineers use copper for the inner liners of combustion chambers, where its ability to quickly channel heat away from hot spots helps prevent localized melting. The copper basically acts like a super-highway for heat, distributing it evenly throughout the cooling system.
- Niobium is a rare metal that remains stable at extremely high temperatures. While less common in everyday life, niobium is invaluable in rocketry, particularly in nozzles where temperatures are most extreme. It maintains its structural integrity in conditions that would turn ordinary steel into liquid, making it perfect for the parts of the engine exposed to the most intense heat.
- Inconel is a nickel-based superalloy that combines the best properties needed for rocket engines. Developed specifically for extreme environments, Inconel maintains tremendous strength even when glowing hot, and resists both oxidation and thermal fatigue. This makes it ideal

for components that must endure both intense heat and enormous mechanical stress simultaneously of chamber walls and nozzles.

The combustion chamber demands absolute perfection in its operation. If the regenerative cooling system falters even briefly, or if the propellants don't mix in exactly the right proportions, or if any tiny component experiences the slightest failure, the consequences are immediately catastrophic. Imagine a household pressure cooker exploding, then multiply that force by thousands—that's the potential destructive power contained within these chambers. The combustion chamber's primary purpose is to safely contain this controlled explosion and channel all that violent energy in a single direction through the next part of the engine that has an equally violent job to do, but perhaps more awesome to witness.

Metallic Tulips

At the tail end of a rocket engine sits one of the most elegantly crafted parts: the nozzle. It might resemble a steel trumpet or, in versions designed for the vacuum of space, a shiny metallic tulip, but it takes this unholy mess of heat, pressure, and explosive chaos, and wields it into focused thrust that propels rockets into space, making these incredible journeys possible.

Modern rocket nozzles trace back to the 1880s and Swedish inventor Gustav de Laval (1845-1913). De Laval designed a special nozzle—now called the de Laval nozzle—while improving steam engines. This nozzle accelerates gases to speeds beyond sound by forcing them through a narrowest point called the throat, then letting them expand into a bell-shaped section. Though originally made for steam turbines, the design proved ideal for rockets.

The defining feature of the de Laval nozzle is its hourglass shape, with the throat as the narrowest section. Hot, high-pressure gases from the combustion chamber enter through a narrowing section, speed up to the speed of sound at the throat (a state called choked flow), and then expand outward. The size of the throat sets the maximum mass flow rate of gas leaving the engine.

Beyond the throat, the nozzle flares outward. In this expanding section, the now supersonic gas spreads out and accelerates further, trading pressure for velocity (remember Bernoulli?) and turning much of its thermal energy into kinetic energy. This produces a rapid, high-velocity exhaust which, by Newton's Third Law of Motion, pushes the rocket in the opposite direction, creating thrust.

De Laval's invention underpins almost every modern liquid rocket, from Germany's V2 to America's Saturn V, whose engines expelled exhaust at roughly two kilometers per second. The principle is the same as partially covering the end of a garden hose to make the stream shoot farther: the narrowing throat boosts pressure and speed, while the flared section finishes the job of turning random push in all directions into an organized shove in one. It is compressible flow physics–those "science class" ideas about pressure, density, and velocity–applied at thousands of degrees and enormous pressures. I am not entirely sure if Euler, Bernoulli, and Newton would have had any idea their new laws of physics would be applied in this way!

The nozzle has to survive the punishment of gas roaring through it, which can reach temperatures hotter than 3,000°C. In the past, early rockets used ablative materials that slowly burned or eroded away over time. Today, many engines use regenerative cooling, similar to how the combustion chamber works: super cold propellant flows through tiny channels in the nozzle wall before entering the chamber. This flow of cryogenic fuel helps prevent the metal from melting and also preheats the propellant, which is a smart way to use the heat instead of fighting it.

Different Types of Nozzles

Early rockets had simple cone-shaped nozzles, but these weren't very efficient–especially in space–because they couldn't guide the exhaust gases to produce the most thrust. Engineers later created the more familiar bell-shaped nozzle, which channels and expands the exhaust more smoothly, helping to reduce wasted energy and boost performance.

Designing a rocket nozzle isn't as straightforward as choosing the perfect shape. It's challenging because a rocket needs to work well under very different atmospheric pressures during its journey. At sea level, where the air is thick and pressure is high, the exhaust must be contained and directed so that its pressure closely matches the surrounding air. That's why a sea-level nozzle is usually short and narrow, with a smaller expansion ratio (the size of the exit compared to the throat). If the exhaust expands too much in these conditions, its pressure drops too low, flow separation can occur, and efficiency drops–sometimes causing damaging instabilities.

Once the rocket reaches space, the situation changes completely. With almost no surrounding air pressure, the exhaust wants to expand much more than it could at sea level. That's where the vacuum nozzle comes in with its larger expansion ratio and a wide exit flare, allowing the gases to expand fully and their pressure to drop close to zero in the vacuum of space. This maximizes how well the heat energy is converted into exhaust velocity, which helps achieve maximum thrust and efficiency in orbit or deep space.

The trade-off is that a nozzle that works perfectly at sea level won't be ideal in space, and one designed only for a vacuum will perform poorly–and could even become unstable–at launch. Engineers balance these competing needs when designing engines that must operate from the ground all the way up to orbit. That's why different rocket stages feature different types of engine nozzles - another benefit of staging we were talking about.

This challenge inspired some innovative solutions. During the 1950s and 60s, engineers came up with the aerospike nozzle–a design that uses a central tapered spike instead of a traditional bell. Hot gases flow along this spike, while the surrounding atmosphere helps shape the exhaust flow. As the rocket climbs and atmospheric pressure drops, the exhaust naturally spreads wider, providing automatic altitude adjustment. This self-adjusting design keeps the engine running efficiently throughout the entire ascent, from liftoff to space.

NASA and several aerospace companies, including Rocketdyne and Lockheed Martin, worked on aerospike engines. The most famous planned use was for the X-33 VentureStar spaceplane, though this project was cancelled in 2001. Today, companies like Firefly Aerospace and ARCA Space are taking a fresh look at aerospike technology, especially for reusable rockets, but none have yet reached operational status on an orbital rocket.

Modern rocket nozzles are made from advanced materials like copper, Inconel, and other special metal alloys designed to survive the extreme heat and stress of space launches. Engineers now use 3D printing to manufacture nozzles with complicated cooling channels built right inside—features that traditional methods simply couldn't create. For example, the SpaceX Crew Dragon's SuperDraco engines use 3D-printed Inconel nozzles with built-in cooling systems to keep them working safely.

Rocket Lab also applies cutting-edge manufacturing to its rocket nozzles. For their Electron rocket, Rocket Lab uses printers to create major engine components—including the Rutherford engine's combustion chamber, injector, and nozzle—through electron beam melting. This advanced 3D printing technique allows Rocket Lab to produce lightweight, strong nozzles with intricate cooling channels and shapes. By printing these parts as solid metal in one piece, Rocket Lab improves reliability and performance, making rapid manufacturing and easy design changes possible.

Whenever you see a rocket launch, remember that the bright flame at the bottom is shaped and guided by some of the most sophisticated engineering ever developed. These nozzles transform intense, explosive energy into the focused thrust needed to send rockets soaring into space—a triumph of technology and innovation.

Thunder Steer

A rocket, for all its fury, is utterly useless without control. You can generate seven million pounds of thrust, shoot fire through a precision-forged or printed nozzle, and vaporize launchpads—and still miss your target by a continent. Raw force is impressive, but directed force is what reaches orbit.

The main steering system for rockets is called a gimbal system. It works like a movable joint that lets the entire engine pivot, tilting the thrust so the rocket turns. When the engine tilts slightly to one side, it changes the direction of thrust, causing the whole rocket to turn. This is similar to how a boat's rudder works—when you turn the rudder in water, it redirects the flow and changes the boat's direction. Similarly, tilting the rocket engine redirects the powerful exhaust gases, guiding the massive vehicle through the atmosphere with remarkable accuracy.

As rocket technology advanced, engineers began adding new steering mechanisms tailored to different phases of flight. One of the most important came from missile technology: the grid fin. Early examples appeared on Soviet designs such as the R400 Oka, and grid fins are now a hallmark of modern reusable rockets, most famously SpaceX's Falcon 9.

These distinctive, square, waffle-like panels excel at providing control during atmospheric flight–especially in the intense conditions of descent and reentry. Unlike conventional aerodynamic surfaces, which work best in smooth, predictable airflow, grid fins perform superbly in the chaotic, highspeed turbulence a rocket experiences when falling back toward Earth, engine first. Their open lattice structure lets air pass through while still generating the forces needed for steering, preventing the stalls or loss of control that solid surfaces might suffer in such conditions.

On the Falcon 9, grid fins are a key part of the landing process. As the rocket returns from space, three engines fire to slow it down, but only the center engine gimbals–pivoting to adjust thrust direction– while the others shut down one by one. Grid fins handle most of the steering in this phase, operating much like the stabilizing feathers on a dart. They keep the vehicle pointed correctly and guide it toward the landing pad or drone ship.

By maintaining steady control authority during the most challenging parts of flight, these simple devices have become a vital part of modern reusable rocketry. They wonderfully show how concepts from missile engineering have contributed to some of the most memorable landings in spaceflight history.

Reaction Control System (RCS) thrusters revolutionized spacecraft steering in the Gemini and Apollo era. The main engines provide the big pushes; RCS thrusters handle the fine work–short, precisely timed bursts that trim a spacecraft's orientation and position.

They usually use hypergolic propellants such as hydrazine and nitrogen tetroxide, which ignite the instant they meet, removing the need for complex igniters and making each pulse highly reliable. Thrusters are arranged in balanced pairs around the vehicle, giving full control of pitch, yaw, roll, and sideways or foreaft translation. In the vacuum of space, where wings and rudders are useless, these small jets become the primary steering system, enabling delicate tasks like docking, aiming solar panels, and making tiny course corrections on interplanetary flights.

All of a rocket's steering systems work in concert, each taking the lead when conditions favor it. Gimbaled engines handle large, forceful movements; grid fins manage control during the chaotic rush of atmospheric flight; and RCS thrusters take over for precise maneuvering in the vacuum of space. The timing of each system is critical: as the rocket climbs and the air thins, aerodynamic controls gradually lose effectiveness, and steering authority shifts increasingly to engine gimbals.

Different rockets use these tools in various combinations. The mighty Saturn V relied entirely on gimbaled engines to keep itself upright. The Space Shuttle combined gimbals with control from its twin solid rocket boosters. SpaceX's upcoming Starship will pair gimbaled engines with massive aerodynamic flaps to execute its distinctive, bellyflop-style landing.

Overseeing all of this is a guidance computer small enough to fit in a suitcase. This system constantly gathers data from sensors, gyroscopes, and GPS, tracking the rocket's position, speed, and orientation. In less than the blink of an eye, it calculates the necessary adjustments—compensating for everything from crosswinds to the sloshing of propellant inside the tanks—and issues commands to each steering system so that the rocket remains on its path.

Moving Forward

Rocket propulsion is built on a straightforward principle from Newton's Third Law of Motion—that for every action, there is an equal and opposite reaction. In a rocket, the "action" is hot exhaust gases being forced out at high speed, and the "reaction" is the rocket itself being pushed forward—producing *thrust*. This principle works just as well in the vacuum of space as it does in Earth's atmosphere, which is why rockets remain our primary method for reaching orbit and traveling beyond.

Launching a rocket, however, is far more complicated than the simple principle might suggest. It's like trying to lift an enormous and constantly changing weight. At liftoff, a rocket is at its heaviest, burdened by its own structure, its payload, and, most of all, the massive quantity of fuel it must carry. As it burns that fuel, the rocket grows lighter, and as it climbs higher, the air thins—gradually reducing drag but altering how the engines interact with the surrounding environment. To leave the ground, the engines must generate enough thrust to overcome Earth's gravity while withstanding the stresses of launch.

Two key measurements guide engineers in meeting this challenge. The first is the thrust-to-weight ratio (TWR), which tells how powerfully a rocket can accelerate compared to its weight. For a launch vehicle, a TWR between about 1.2 and 1.5 is typical. Any lower and the rocket would struggle upward against gravity; any higher and the excess power could damage the structure or waste precious fuel. The second is the specific impulse (Isp) that we learned about before that measures how efficiently an engine uses its fuel. Conventional chemical rockets burning liquid hydrogen and oxygen can achieve around 450 seconds of Isp, while solid-fuel rockets deliver 250 to 300 seconds, trading some efficiency for raw thrust. At the other extreme, ion engines can achieve an impressive 3,000 seconds or more—but produce very low thrust, making them unsuitable for liftoff.

This is why spacecraft propulsion is a balancing act. Chemical rockets are the sprinters of spaceflight, delivering the brute force needed to escape Earth's gravity well but burning through fuel quickly. Ion engines are the marathoners, providing a steady, ultraefficient push for months or even years—ideal for deepspace travel where patience and efficiency outweigh raw power.

Modern rocket designers must weigh all of these factors: the mission goals, the choice of fuel, storage requirements, the staging of engines, and how performance can be maintained from launch pad to final destination. In the end, it's an extraordinary achievement—transforming stored chemical or electrical energy into highly controlled motion across the vastness of space. It's a feat made possible by centuries of progress in physics, thermodynamics, fluid mechanics, and material science, culminating in one of humanity's most sophisticated engineering triumphs.

Chapter 18

Other Types of Rockets

Horses for Courses

So far, we've explored the basic principles of liquid-fueled rocket engines and their key components like nozzles and control systems. However, this is just one approach to rocket propulsion. The field of rocket science has developed numerous engine designs, each trying to overcome specific challenges in unique ways.

Different missions require different solutions. While some rockets use the traditional combination of liquid hydrogen and oxygen, others employ solid fuels packed into combustible cores, or hybrid systems that combine both approaches. Some advanced designs even abandon chemical propulsion entirely, using electric fields to accelerate ions to incredible velocities. We also have specialized engines like aerospikes that maintain efficiency across different atmospheric pressures, and designs like SABRE that can switch between air-breathing and rocket modes.

Each of these engine types are different solutions to our fundamental challenge: how to generate thrust as efficiently as possible for specific mission requirements. In the following chapters, we'll examine each type in detail—their working principles, construction materials, advantages, and the new challenges they introduce.

Remember that every rocket engine design involves trade-offs between performance, complexity, and reliability. While each approach has its limitations, these various innovations have not only advanced space exploration but also contributed valuable technologies to other fields.

Self-Consuming Rockets

The simplest type of rocket engine is the solid rocket—a design that trades the complex plumbing of liquid-fueled engines for straightforward, rugged simplicity. Instead of separate tanks, pumps, and valves, a solid rocket is a strong tube packed with a solid propellant: a premixed blend of fuel and oxidizer bound together.

Despite appearances, these rockets don't erupt all at once like giant firecrackers. Their burn is carefully controlled, beginning at the core and progressing outward. At the heart of this precision is *grain geometry*—the engineered shape of the propellant inside. By sculpting this geometry, engineers can dictate how the burn surface grows over time, shaping the thrust profile throughout the rocket's flight. While a solid rocket's thrust and burn time can be predicted very accurately, it cannot be throttled or shut down once it has been lit.

These qualities are why solid rockets excel at one of spaceflight's toughest jobs—liftoff. The Space Shuttle, for example, depended on two enormous solid rocket boosters that produced searing exhaust reaching temperatures of thousands of degrees, hotter than many industrial furnaces, helping the orbiter muscle its way through the thickest part of the atmosphere. Once spent, the empty boosters detached and dropped into the ocean, ready to be recovered and refurbished.

Even today, solid rockets remain a cornerstone of aerospace engineering. Their simplicity, reliability, and rapid readiness make them ideal for military missiles, emergency escape systems, and quick-launch vehicles. Beyond space and defense, the same technology has even found work here on Earth—in mining, demolition, and other industries where controlled bursts of immense power are needed.

From War Rockets to Wedding Sparklers

The history of solid rockets stretches back centuries—long before they were used to send astronauts into space, they were lighting up the skies in entirely different ways. As we mentioned earlier, back in thirteenth-century China, early inventors packed gunpowder into bamboo tubes to create "fire arrows," representing the earliest form of primitive rockets. Over time, these weapons evolved into stunning fireworks that add vibrant colors and lively sounds to our celebrations.

For many years, rockets were quite unpredictable. Some flew straight, while others veered off wildly, and many fizzled out before they even launched. But things started to change in the twentieth century, when dedicated military and space agencies began studying rockets in earnest. During World War II and the Cold War, pioneers like Jack Parsons and engineering teams at JPL, Thiokol, and the US Navy made incredible breakthroughs. They improved fuel mixtures, shaped propellants to control burn rates, and designed ignition systems that worked flawlessly every time. Although these innovations were initially aimed at missiles and spacecraft, they quickly found their way into many other areas of our daily lives.

Thanks to rocket research, modern fireworks are safer, more reliable, and more precise. Today's shells fly straight, ignite on cue, and explode into highly choreographed colors. Even everyday items like sparklers on the Fourth of July or during Britain's Guy Fawkes Night are made from refined solid-propellant compositions originally perfected for rockets. The same technology is used for emergency flares, ensuring dependable ignition in any weather and long storage without losing effectiveness. In space missions, similar charges help separate rocket stages or deploy parachutes—tasks where absolute reliability is essential.

Surprisingly, rocket science has even helped save lives on the road. In the 1970s, car manufacturers looked to space-grade technology to develop faster, lighter airbags. Instead of heavy compressed-gas tanks, they adapted small solid-propellant charges initially used to jettison rocket parts. When sensors detect a crash, a tiny rocket ignites, releasing nitrogen gas that inflates the airbag in just twenty to thirty milliseconds—faster than a blink of an eye. By the 1980s, these rocket-powered airbags were common in cars, and by the late '90s, they became standard around the world.

Deep underground, rocket technology is also making a difference. In the 1950s, great advances were made in controlling solid-fuel combustion, strengthening casings, and making ignition more reliable. Mining and construction industries quickly adopted these innovations. While traditional explosives were powerful, they lacked precision. By using rocket-derived detonators with microsecond timing, engineers gained incredible control over blasts, shaping explosions exactly where and when needed. Materials designed to withstand launch stresses found new uses in tough drilling tools, and the complex mathematics used to predict rocket trajectories was reimagined to model how rock would fracture using carefully timed charges.

Today, the same science that started with bamboo "fire arrows" now brightens our skies, keeps us safe on the roads, and even helps shape the ground beneath our feet. Whether it's a flare that could save a sailor's life, an airbag that protects a driver in a crash, or the controlled demolition of a skyscraper in a busy city, all share a common origin: centuries of learning to understand and harness the power of fire.

This is the kind of rocket science we like to see blow up!

Pyrotechnic Life Savers

Hybrid rockets sit at the crossroads between the two main types of rocket propulsion. Solid rockets are like giant fireworks—they burn all their fuel once lit and can't be turned off—while liquid-fueled rockets are complex, highly controllable machines that require intricate plumbing and careful handling. A hybrid rocket blends aspects of both, pairing a solid fuel—often a rubbery block—with a liquid oxidizer such as nitrous oxide or liquid oxygen. The result is a system that delivers much of the power of a solid motor but with the controllability of a liquid engine.

Although scientists had been toying with hybrid rocket concepts since the early 1900s, the technology didn't really advance until the Cold War, when there was demand for rockets that were cheaper, potentially safer, and easier to handle than many traditional liquid systems. They proved useful for research vehicles, certain missiles, and, later, experimental spacecraft.

A landmark came in 2004, when SpaceShipOne became the first privately built craft to reach space using a hybrid engine. Its design was simple and used some surprising materials: HTPB (*Hydroxylterminated Polybutadiene*, a synthetic rubber similar to what's found in car tires) as the solid fuel, and nitrous oxide–better known as "laughing gas"–as the oxidizer. The rubber fuel grain was cast with a hollow core running the length of the engine. During a burn, liquid nitrous oxide was sprayed into this cavity and ignited; the intense heat vaporized the rubber, which then burned in a controlled manner, though engineers still had to carefully manage how smoothly the fuel burned and how evenly it mixed.

Unlike traditional liquid engines, hybrids don't need cryogenically cold, dangerous propellants, and they're much simpler than complicated liquid systems. Unlike solid rockets, they can be started, stopped, and even throttled mid-flight. This makes them accessible for building and testing in regular workshops and hangars, which is great news for universities, private groups, and smaller companies that previously couldn't afford to get into orbital technology. However, scaling hybrids to very large, smooth-burning engines has been tricky, which is why most big launchers still stick with pure liquid or solid engines.

The principle is simple: take a cylinder of solid fuel, inject a liquid oxidizer, and you have a controllable burn–an enormous advantage over solid rockets, which keep firing until their propellant is gone. In practice, engineers still have to manage how fast the fuel surface erodes and how smoothly it mixes.

Modern fire suppression systems in sensitive environments–like nuclear submarines or server rooms–highlight how this mastery of controlled combustion and gas flow has extended beyond rockets. Instead of using water, which could damage electronics, these systems store inert gases and special solid compounds. When needed, precise metering systems–drawing on the same kinds of valves and flow controls originally developed for rocket oxidizers– quickly flood the room with a balanced gas mixture that extinguishes flames safely without harming equipment. Since their introduction in the 1980s, systems like INERGEN and FM-200 have helped prevent billions of dollars in potential damages.

Because hybrid rockets are easy to store and handle under normal conditions, they've sparked a wonderful boom in amateur and budget-friendly rocketry since the 1990s. This has opened doors for many enthusiasts to get involved in serious propulsion projects. For example, groups like Copenhagen Suborbitals have used hybrid engines to create impressive spacecraft on modest budgets, inspiring students, hobbyists, and small teams to follow their dreams of spaceflight.

The same careful engineering of valves and flow control—once used to meter oxidizer into hybrid rocket chambers—now also helps in hospital operating rooms, where it ensures precise regulation of anesthesia gases in milliseconds. Likewise, in the food industry, these systems keep gas mixes inside packaging perfectly controlled to maintain freshness, and breweries use them to fine-tune carbonation, so each pint always tastes just right.

From thirteenth-century "fire arrows" to today's private spacecraft and advanced operating rooms, hybrid rocket technology shows us how mastering controlled combustion and gas flow—whether in the depths of space or the calm of a hospital—can transform our lives far beyond the launch pad.

Sometimes, the most amazing part of rocket science is making sure nothing explodes at all.

Electron Breathers

At first glance, ion engines seem underwhelming—their thrust is so slight that a feather on your palm exerts more force. Yet in space's vacuum, these patient engines eventually outperform chemical rockets in total speed change, given enough time. Where chemical rockets are hard-burning sprinters, ion propulsion is the marathon runner—slow, steady, and remarkably efficient. In deep space's emptiness, this tortoise-over-hare approach wins decisively.

Ion engines rank among the most efficient propulsion systems available. Rather than using explosive chemical reactions, they employ electricity to accelerate ions to extraordinary speeds. Like a garden hose that delivers a continuous stream instead of a fire hose's brief blast, ion engines produce modest but unrelenting thrust that can operate for years on minimal fuel. This sustained acceleration ultimately allows spacecraft to reach velocities beyond what chemical rockets alone can provide.

Our friend, Konstantin Tsiolkovsky who also gave us the Rocket Equation, first conceived this idea. He envisioned propelling spacecraft with electrically charged particles instead of combustion—revolutionary thinking for his era. Decades later, this concept took practical form in 1959 when NASA's Harold Kaufman built the first working ion engine at Glenn Research Center, using electromagnetic fields to accelerate ionized gas like a microscopic particle accelerator.

NASA's early flight tests, Space Electric Rocket Test (SERT-I) in 1964 and SERT-II in 1970, ran ion engines in space long enough to prove that this once-speculative idea could actually work.

Despite these early successes, ion engines remained experimental for decades because chemical rockets delivered the immediate thrust needed for launches. However, ion propulsion offered an important benefit: it could provide continuous thrust for extremely long durations while consuming minimal propellant. By the 1990s, NASA was ready to harness this capability. The watershed moment came in 1998 with Deep Space 1, the first spacecraft using an ion engine as its primary propulsion. Operating for 16,000 hours—like an engine running continuously for nearly two years—it showed that ion propulsion could power long duration missions reliably. That success led directly to the Dawn mission (2007), which visited Vesta and Ceres using three ion engines that operated for over 25,700 hours during its eleven-year journey—an endurance record that would be impossible with conventional rockets.

Today, ion propulsion is woven into many parts of spaceflight. Earth-orbiting satellites use ion engines like precision rudders, maintaining their orbits and making careful adjustments while using far less propellant than traditional thrusters. This thriftiness extends satellite lifetimes well beyond the usual span and even enables "all-electric" satellites that rely on ion propulsion for most of their orbital maneuvers. Miniaturized ion thrusters have also changed CubeSats from passive drifters into nimble explorers capable of changing orbits, dodging debris, and flying complex formation patterns.

Looking ahead, ion propulsion will play a crucial role in lunar and Mars exploration, asteroid mining, and maintaining future space habitats—becoming a cornerstone of deep-space travel and sustainable long-duration missions.

Newton Meets Plasma Physics

Ion engines are rockets that swap flames for electricity. Instead of burning fuel, they use electric fields to push charged atoms out of the back of the spacecraft, so the ship moves forward in the opposite direction, just as Newton's Third Law demands: every action produces an equal and opposite reaction.

Xenon gas is used as the propellant for three main reasons. First, xenon atoms are heavy, so when they are accelerated and expelled from the thruster, each ion carries a lot of momentum—more like throwing a bowling ball than a pingpong ball. Second, xenon is also a noble, chemically inert gas, so it does not corrode engine parts and can sit safely in tanks for years, which is vital for long deepspace missions. Finally, xenon is monatomic: its atoms travel solo, not paired into molecules. That means the thruster does not waste energy breaking molecular bonds before ionization; the electric field can go straight to ionizing and accelerating individual atoms, making the process more efficient.

Inside the thruster, xenon gas flows into a small chamber where energetic electrons knock electrons off the xenon atoms, turning them into positively charged ions and creating a cloud of plasma. These ions are then pulled through a region of very strong electric field—typically hundreds to more than a thousand volts—which flings them out of the engine at tens of kilometers per second, like microscopic bullets leaving an electric cannon. The force from each ion is tiny, far less than the weight of a sheet of paper on your hand, but because the engine can run steadily for weeks or months, the small push keeps adding up. Over time, this continuous acceleration lets spacecraft reach higher speeds than a chemical rocket can after its brief, powerful burn is finished.

If the engine ejected only positive ions, the spacecraft would quickly build up a negative charge and start pulling its own exhaust back, cancelling the thrust. To prevent this, a separate cathode injects electrons into the exhaust plume so the mix of ions and electrons remains electrically neutral and can drift away into space without being pulled back toward the spacecraft. This neutralization step is essential for ion propulsion to work over long durations without charging up the spacecraft.

Because they use propellant so efficiently, ion engines are ideal for jobs that demand precise, long-lasting thrust rather than brute force. Modern satellites use them to make gentle, continuous orbital adjustments, keeping their paths tuned and helping them dodge debris while consuming only a small amount of xenon. For deepspace probes, the same gentle push can operate for years, turning a modest amount of propellant into enormous total speed—like compound interest slowly turning a small deposit into a large balance over time.

A New Way of Thinking

Ion engine research started in the late 1950s for spacecraft, and it's part of a wider wave of plasma technology that's transformed life on Earth behind the scenes. In the 1970s, improvements in low-temperature plasma—using many of the same physics and engineering ideas as ion propulsion—revolutionized medical sterilization. Instead of relying on scalding steam that could harm delicate instruments, plasma-based systems now use reactive ions and radicals at much lower temperatures, which can clean tools in less than an hour, save energy, and reduce the need for harsh chemicals.

Plasma and ion beam techniques also revolutionized how we make microchips. In modern semiconductor fabs, tightly focused ion beams and reactive plasmas "etch" ultrafine patterns into silicon, like atomic-scale paintbrushes tracing circuits on canvases smaller than dust motes. This level of control lets engineers pack billions of transistors onto fingernail-sized chips, which in turn power everything from smartphones in our pockets to laptops on our desks.

Protecting ion thrusters from the tough space environment also led to improvements in ultra-thin, durable coatings. Engineers developed ceramic and glass-like layers that can withstand energetic particles and erosion. These coatings now help make smartphone screens tougher, reduce glare on camera lenses, extend the lifespan of solar panels, and improve how well medical implants get along with the human body. Even though these innovations come from fields like materials science, optics, and aerospace, work on electric propulsion has really helped advance these technologies.

Power management is another important link between space tech and everyday life. The advanced control systems designed to deliver steady, efficient power to ion engines are quite similar to those used in modern electric cars: both need to get the most out of limited energy, regulate voltage and current, and protect sensitive parts. Lessons learned in one area often help improve the other, promoting a design approach that values efficiency and precise control over simply applying raw power force.

Looking Ahead

In space, ion propulsion is more than just an impressive engineering achievement—it's a crucial step in our journey to explore other planets. Soon, efficient ion-powered ships will ferry people and cargo between Earth, Mars, and the asteroid belt, making space travel easier and more accessible than ever before. While chemical rockets will still give us the initial boost to escape Earth's atmosphere, once in space, these quiet electric thrusters will gently steer ships through the universe with impressive efficiency—much like switching from gas-guzzling highway driving to smooth, eco-friendly electric cruising.

Their proven reliability makes them perfect for a variety of exciting missions: supporting long-term orbital stations, maintaining lunar bases, aiding asteroid mining efforts, and sending probes to explore Jupiter or Saturn's icy moons. More than anything, they reflect qualities that are truly part of human progress—patience, ingenuity, and unwavering determination.

Though ion engines might not make loud roars from launch pads, they are ready to lead us into our next chapter of exploration - confidently, and limitless in potential.

Atomic Propulsion

During the Cold War's Space Race, engineers developed nuclear thermal rocket (NTR) engines—rockets that used the heat from a nuclear reactor to supercharge their propellant instead of relying solely on chemical combustion. Liquid hydrogen was pumped through a uranium-packed reactor core, where nuclear fission heated it to 2,500-3,000°C before expelling it as high-velocity exhaust. Because nuclear reactions pack far more energy into a given mass than chemical reactions, these engines could roughly double the efficiency of chemical rockets, letting spacecraft travel farther with less propellant—a tempting prospect for future Mars missions.

Two major programs pushed this technology forward: Project Rover (1955–1972) focused on developing the reactors themselves, with the 1959 KIWI-A test proving fission could effectively heat propellant for space propulsion. NASA's follow-on NERVA program (1961-1973) then built full-scale engines, producing the XE-Prime that delivered 75,000 pounds of thrust while maintaining roughly twice the efficiency of chemical rockets. By 1969, the XE-Prime could be repeatedly started, throttled, and shut down, operating for over an hour under simulated space conditions.

Despite these technical successes, the program was cancelled in 1973. After the Apollo 11 Moon landing, public enthusiasm for expensive space projects faded, while Cold War nuclear safety concerns and shifting space policy priorities led to funding cuts. Although no nuclear thermal rocket has yet flown in space, the concept is being revived as interest in crewed Mars missions grows—more than a half century after those first Nevada ground tests.

Benefits on Earth

Even when there isn't a flight mission, the work on nuclear thermal rockets has given us some very impressive advances. Scientists have developed high-temperature materials like carbide-graphite composites, tungsten-rhenium alloys, and zirconium-based ceramics that can handle incredible heat—over 2,500°C. These innovative materials, along with their newer versions, are now found in jet-engine turbines, combustion liners, and industrial furnaces reaching nearly 2,000°C. They're playing a big role in saving energy compared to older designs. Plus, power plants are also benefiting from advanced superalloys and ceramic parts, helping us produce more electricity from each unit of fuel.

Ideas from compact reactor designs initially meant for Mars transfer stages have inspired small nuclear power systems that serve remote military bases and research stations in tough environments—places where traditional power grids aren't practical. Similar concepts could soon power disaster response units and support new outposts in isolated areas on Earth and beyond.

Research into radiation, initially linked to human spaceflight, has greatly benefited modern medicine as well. The methods developed to protect astronauts and precisely calculate radiation doses have helped advance cancer treatments. Today, tools like multileaf collimators and other beam-shaping devices enable us to target tumors with incredible accuracy—sometimes down to less than a millimeter. This means more effective treatments with fewer side effects. Shielding and careful dose management keep both patients and medical staff safer during diagnostic procedures. For instance, the newest PET-CT systems greatly cut down occupational exposure compared to older models.

Building on this legacy, NASA and Defense Advanced Research Projects Agency (DARPA) are collaborating on the Demonstration Rocket for Agile Cislunar Operations (DRACO1), scheduled for testing between 2026 and 2027. This next-generation nuclear thermal rocket will use a compact reactor to superheat hydrogen fuel, potentially reducing travel times between Earth and the Moon while increasing payload capacity beyond chemical rockets' capabilities.

Other efforts include Project Pele, which is developing transportable microreactors in the megawatt range for future bases, and NASA's Kilopower project, which focuses on kilowattscale reactors to support long-term human missions on the Moon or Mars. Earlier initiatives such as Project Prometheus helped lay groundwork for nuclear electric propulsion—systems that would use a reactor to power high-efficiency electric thrusters for deepspace journeys.

The pioneering work of the NERVA program proved that nuclear propulsion could surpass chemical rockets in efficiency and power, and its spirit lives on. As humanity prepares to send astronauts to Mars and beyond, there is a real possibility that their transfer stages will be advanced descendants of those Nevada test engines, updated with half a century of progress in materials, safety, and reactor technology.

Quiet Engine

In a world filled with noisy combustion engines, the Stirling engine shines as an impressive wonder of engineering. This special machine offers a completely different way of generating power, which is becoming more and more important for both space exploration and sustainable energy efforts here on Earth.

The story of the Stirling engine begins in 1816 with Reverend Robert Stirling (1790–1878), a Scottish minister and inventor. During a period when steam engines were often plagued by dangerous boiler explosions that tragically took many lives in factories and ships, Stirling was determined to find a safer solution. His innovation, known as the "Heat Economizer," offered a new, safer way to make use of heat energy, sidestepping the hazards linked to high-pressure steam.

The Stirling engine operates on a wonderfully simple, closed-cycle principle that's both elegant and effective. Instead of burning fuel inside the engine like traditional combustion engines, it contains a fixed amount of gas—such as helium, hydrogen, or air—sealed inside. This gas moves between two chambers: one heated and one cooled. When the gas warms up in the hot chamber, it absorbs heat and expands, pushing a piston to produce mechanical work. Then, as it enters the cooler chamber, it contracts as it cools, pulling the piston back. This ongoing cycle of expanding and contracting turns heat into motion smoothly, efficiently, and almost silently.

What makes the Stirling engine so useful is its incredible ability to adapt to different heat sources. It can operate with almost any temperature difference, whether it's from burning fuel, concentrated sunlight, geothermal heat, or even nuclear decay. This versatility has attracted the interest of space agencies and renewable energy enthusiasts everywhere, highlighting its potential for a wide range of innovative applications.

Since the 1970s, NASA's interest in Stirling technology has grown a lot, especially for deep-space missions where solar panels and batteries become less effective. A major breakthrough has been the development of Advanced Stirling Radioisotope Generators (ASRGs). These combine Stirling engines with plutonium-238 radioactive heat sources and can produce over twice the efficiency of traditional thermoelectric generators. This means spacecraft can operate longer while using less nuclear fuel–an especially important benefit given the limited availability of space-grade plutonium.

Although NASA cancelled the ASRG program in 2013 over reliability and funding concerns, research continues, and Stirling-based space power remains a promising technology for future long-duration missions, such as Mars rovers and lunar bases. Modern ASRGs can maintain reliable power output for decades, showcasing the potential for quiet, efficient energy in challenging environments.

On our planet, Stirling engines are enjoying a wonderful resurgence in renewable energy and local power generation. Solar-Stirling plants, with their large mirrors focusing sunlight, heat these engines, transforming solar energy into electricity with some of the highest efficiencies ever seen. Thanks to their low maintenance and ability to work with various heat sources, they are perfect for remote areas and backup power needs.

Thanks to recent advancements, Stirling technology has also helped develop small-scale micro-CHP units that provide both electricity and warmth for homes and businesses, boosting energy efficiency and sustainability.

Looking ahead, Stirling engines are set to play an important part in building sustainable energy systems around the world. Their reliability, quiet operation, and fuel flexibility make them ideal for powering everything from off-grid communities to space stations. Kind of cool how a device invented over two-hundred years ago could be leading a new energy revolution for the twenty-first century!

Air Breathers

Imagine an engine that combines the best qualities of jet engines and rockets. The Synergetic Air-Breathing Rocket Engine (SABRE), developed by Reaction Engines Ltd in the UK since 1989, is being designed as a breakthrough in aerospace propulsion technology. Building on concepts from the 1980s HOTOL project and informed by decades of hypersonic flight research–including the 1960s Franco-German ISOS project and NASA's X-30 National Aerospace

Plane program—SABRE aims to transform space travel by reducing the weight and complexity of reaching orbit.

What makes SABRE unique is its dual-mode operation. In the atmosphere, it acts like an advanced jet engine by breathing air to supply oxygen for combustion, eliminating the need to carry heavy onboard oxygen tanks during the early part of flight. In this air-breathing mode, the engine is intended to operate efficiently up to hypersonic speeds of about Mach 5 at altitudes near twenty-five kilometers. Beyond this, as the atmosphere thins, SABRE would switch to rocket mode, using stored liquid hydrogen and oxygen to accelerate toward speeds above Mach 25—fast enough for orbit.

The key enabling technology behind SABRE's air-breathing capability is its pre-cooler system. At hypersonic speeds, the air rushing into the SABRE engine is moving so fast that when it's suddenly slowed and squeezed inside the intake, all that kinetic energy is turned into heat—a process called stagnation. This causes the temperature of the incoming air to shoot up, sometimes reaching over 1,000°C, which is hot enough to melt metal and destroy engine parts in an instant.

To keep the engine safe, SABRE uses a pre-cooler system made of thousands of tiny tubes. As the super-hot air enters, it passes through these tubes and is quickly chilled using a special coolant, dropping its temperature from over 1,000°C down to as low as -150°C—faster than the blink of an eye. By cooling the air so rapidly, SABRE protects the engine and makes the air dense enough to be easily compressed and burned for thrust, all without risking overheating. This approach lets SABRE use "stagnation" air—the hottest possible form—as part of its high-speed, air-breathing strategy, something earlier concepts struggled to achieve.

In performance terms, design studies suggest that SABRE could deliver high thrust with excellent efficiency. In airbreathing mode, it is projected to produce roughly 1,400–1,500 kN of thrust with a peak specific impulse far higher than that of conventional chemical rockets. In rocket mode, the planned thrust remains in a similar range, with efficiency comparable to top-tier hydrogen–oxygen engines. Its expected thrust-to-weight ratio is significantly higher than that of conventional jet engines and scramjets, helped by the dense, cooled intake air reducing compression work.

The potential of such an engine is very exciting if these goals are achieved. By not needing large oxygen tanks for the ascent, spacecraft could become much lighter, which would boost payload capacity and make single-stage-to-orbit (SSTO) vehicles more feasible than traditional designs. This concept also paves the way for runway-style takeoff and landing for spaceplanes, much like airplanes, making launch and recovery simpler and more familiar.

Beyond just space access, SABRE-type technology could revolutionize long-distance air travel. Some ideas even suggest a spaceplane could fly from London to Sydney in about four hours, using sustained hypersonic air-breathing flight to drastically reduce travel times around the globe.

Reaction Engines has shared positive updates, including successfully ground-testing the pre-cooler at conditions similar to around Mach 5. They've gained support from the European Space Agency, the UK Space Agency, and major aerospace companies like Boeing and Rolls-Royce. While SABRE is still in development and much work remains before a full engine test, if it approaches its design goals, it could make space access more routine, flexible, and affordable—unlocking exciting new opportunities for space tourism, satellite deployment, and ultra-fast global travel.

Cold Gas Thrusters

Who doesn't like to blow up a balloon and then let it go as it flies all over the room, making a hilarious sound? Well, in the rocket business, we have cold-gas thrusters that kind of work on the same principle.

Cold-gas thrusters represent the elegant simplicity of rocket science at its finest. These devices consist of a few key components: a pressurized tank holding an inert gas such as nitrogen, helium, or carbon dioxide; valves that regulate the gas flow; and specially designed nozzles that direct the escaping gas to produce controlled thrust. When a valve opens, the high-pressure gas rushes through the nozzle, accelerating and generating force in the opposite direction.

Cold-gas thrusters first emerged early in human spaceflight. Although the Mercury program primarily used hydrogen peroxide thrusters for attitude control, it also incorporated small nitrogen cold-gas thrusters as part of the backup system, enabling astronauts like John Glenn and Alan Shepard to make very fine orientation adjustments. Over time, this simple technology gained wider adoption in Gemini, Apollo, and numerous satellite missions thanks to its exceptional reliability. With no combustion, extreme temperatures, or complex moving parts, cold-gas thrusters delivered steady and predictable performance during a pioneering era of space exploration.

Today, cold-gas thrusters remain important for modern spacecraft, especially within the exploding domain of small satellites like CubeSats. Many of these tiny craft, often no larger than a loaf of bread, rely on cold-gas systems for attitude control and station-keeping, benefiting tremendously from the thrusters' compactness, light weight, and simplicity.

The International Space Station also employs cold-gas technology in its Simplified Aid For EVA Rescue (SAFER) units—compact jetpacks astronauts wear during spacewalks. These nitrogen thrusters provide emergency maneuverability, enabling astronauts to safely navigate back to the station if they become untethered—a literal lifeline in the vacuum of space.

Beyond space, cold-gas thrusters have found surprising applications. Some Hollywood special effects teams use similar systems to generate controlled bursts of movement and explosions on film sets. Theme parks incorporate them into rides for smooth, precise motions. In manufacturing, cold-gas jets deliver clean, dry air bursts to remove microscopic contaminants from sensitive electronics and optical components, for making computer chips and high-quality lenses.

What really makes cold-gas thrusters stand out is their impressive combination of simplicity and effectiveness. They shine in situations where reliability is paramount, and operational conditions forbid complex or high-risk systems. While they may not produce as much thrust as chemical rockets or have the specific impulse of electric propulsion, their low chances of failure make them incredibly valuable in critical environments like space.

Engineers continue to refine cold-gas thruster technology. Advances include optimizing nozzle shapes to maximize thrust efficiency, developing stronger and lighter pressure vessels to safely contain gas at higher pressures, and designing more sophisticated control systems enabling ever finer maneuvering precision.

Sustainability is also driving innovation. Traditional cold-gas propellants can have significant global warming potential, so newer systems increasingly use more environmentally benign gases such as compressed air or pure nitrogen. Researchers are even exploring concepts that would harvest and compress trace atmospheric gases in low Earth orbit to create partially self-replenishing propulsion systems—a promising but still experimental idea that could greatly extend mission durations.

Cold-gas thrusters are like a trusted family recipe—simple, reliable, and just right for their special purpose. Their long-lasting legacy, from pioneering human spaceflight to helping grow the small satellite market and ensuring spacewalk safety, shows how the beauty of simplicity can lead to amazing results even in the most challenging environment known to humanity.

It Is Rocket Science

Rocket science is where the seemingly impossible becomes possible. It's a fascinating world where liquids, gases, and metals are driven from freezing temperatures to blazing heat and crushing pressures—sometimes in just fractions of a second. It's where mighty machinery roars to life, flying at speeds twenty-five times faster than sound, and where each material, every bolt, must withstand forces strong enough to tear tanks apart.

In this incredible realm, physics, chemistry, and engineering are stretched to their limits—yet everything must be crafted with perfect precision. A tiny flaw or tiny miscalculation can turn a simple launch into a spectacular event of a very different kind.

From Newton's laws of motion to Tsiolkovsky's Rocket Equation, from Robert Goddard's first liquid-fueled rocket to Wernher von Braun's powerful Saturn V, we stand on the shoulders of giants. They provided us with fundamental principles, and we've pushed them to extremes they could scarcely imagine: materials that endure from near absolute zero to thousands of degrees Celsius, engines that handle pressures once capable of exploding early test chambers to pieces, and spacecraft that navigate the cosmos at unimaginable speeds.

The pioneers of thermodynamics—Carnot, Clausius, Lord Kelvin—would be amazed to see their theories come alive in engines that unleash controlled fury. Maxwell and Boltzmann, leaders in gas theory, would marvel at how we tame the chaotic movement of molecules into the precise flows that power rockets.

Even Newton and Kepler, whose equations still guide spacecraft with incredible accuracy, would be proud to see their work helping us reach for the stars.

Perhaps what would impress them most is how the tools developed for space travel have changed life on Earth. Rocket technology has given us artificial hearts, advanced medical scanners, cleaner energy systems, and super-strong alloys and ceramics. Innovations meant to prevent rockets from melting or shaking apart have led to better turbines, more efficient medical devices, and discoveries in new materials.

Despite jokes about it not being "rocket science," the truth is that rocket science teaches us more than just how to fly—it shows us how to solve some of our toughest problems. It's a wonderful proving ground where humanity's biggest challenges meet incredible ingenuity. And the journey isn't over. Every new launch, every new mission pushes the limits of what we can achieve—not only in space but right here on Earth.

Rocket science reminds us that the lessons we learn chasing the stars will always come back home, helping shape a future that's faster, cleaner, stronger, and more extraordinary than we've ever dared to imagine.

SECTION 3: ORBITING EARTH

Where the Real Space Revolution Happened

While rockets and moon landings undoubtedly caught our imagination, there's an incredible change taking place high above us—around 500 kilometers up—where satellites silently circle our planet. Their quiet presence makes a real impact on our daily lives, shaping our world in ways we might not always see but definitely experience.

Satellites are the backbone of our modern world. They collect data for everything from weather forecasts and agriculture to disaster response and environmental protection. They warn us of storms long before we see them, guide ships across oceans, protect endangered forests, and spot threats before they become crises. They connect the remotest corners of Earth, powering GPS navigation, mobile communications, and internet access—tools we now take for granted.

Their reach goes even further, connecting the world in incredible ways. Financial markets depend on satellite data for quick decisions and to timestamp trades—traders watch satellite images to see parking lots at retail stores, estimate crop yields, track ships, and predict commodity prices. This real-time information guides investment choices and helps keep our global economies stable. At the same time, satellites boost security by watching borders, monitoring natural disasters, and supporting emergency communications when ground networks are down. All these orbiting sentinels work together like the Earth's nervous system, maintaining global order with remarkable precision.

In this section, we'll explore how satellites work, from the sensors that transform light and signals into actionable knowledge to their evolution from Cold War tools to indispensable humanitarian assets. We'll also dive into orbital mechanics that keep these machines circling, and trace their history, revealing how this unseen network became our planet's most powerful eye in the sky.

With this perspective, the greatest achievement of spaceflight isn't about leaving Earth per se—it's about seeing it more clearly than ever before.

Chapter 19

From Sputnik to Starlink

Gazing Back Down

When humans first ventured into space, one of the most unforgettable moments was gazing back at Earth from above. This simple act offered us a new way of seeing—our planet as a small, stunning blue sphere gently floating in the endless expanse of space.

Viewing Earth from this vantage point profoundly changed how we see our home. It helped scientists understand how deeply connected and delicate our planet is, and it inspired people all around the world to cherish and protect this precious place we all share.

That decision to look back transformed our understanding and appreciation of Earth.

The Beep That Started It All (1950s)

The space age began with an unexpected "beep" that echoed around the world. On October 4, 1957, the Soviet Union launched Sputnik 1—a basketball-sized sphere that became Earth's first artificial satellite. It's simple, rhythmic radio signals sent shockwaves across the globe. Amateur radio operators everywhere shared the experience of tuning in to Sputnik's distinctive pulse, while newspapers buzzed with headlines about this new "moon" in orbit. The American public, caught off guard, realized they had lost their technological edge in space. Though Sputnik transmitted signals for just three weeks, its impact was profound—it altered the course of history, triggering the intense space race between the United States and the Soviet Union and sparking massive investments in science and technology education.

In response, the United States launched Explorer 1 in 1958, marking its entry into the space race. This groundbreaking mission led to the discovery of the Van Allen radiation belts—two invisible, doughnut-shaped bands of high-energy particles trapped by Earth's magnetic field. These belts resemble shimmering protective bands encircling the planet, similar to Saturn's rings but composed of energetic particles rather than ice and rock.

Later that year, the US reached another milestone with Signal Communication by Orbiting Relay Equipment (SCORE), the very first satellite designed specifically for communications. It made history by broadcasting a recorded Christmas greeting from President Eisenhower—the first human voice transmitted from space to Earth. This was the first real time that we could see how satellites could bring people closer over great distances in a brand new way.

The First Dial Tone from Space (1960s)

The 1960s brought the first of many major breakthroughs in satellite technology, starting with Echo 1, a giant metallic beach ball in space, one hundred feet across and weighing about 400 pounds. This passive satellite worked like a giant mirror that simply bounced radio signals across continents. It was visible from Earth with the naked eye and did not need any electronics to function.

By 1962, NASA launched Telstar 1—the first active communication satellite that looked like and was the size of a soccer ball. Weighing in at 170 pounds, it featured a distinctive black-and-white pattern of solar cells, amplifiers, and repeaters arranged in pentagons and hexagons. Unlike Echo 1's simple reflection, Telstar was like a space-based amplifier, receiving signals, boosting them, and retransmitting them back to Earth. It could handle 600 voice calls or one TV channel.

Telstar's orbit resembled an egg-shaped path, racing between 593 and 3,503 miles above Earth every 2.5 hours. Ground stations had to track it like hunters following fast-moving prey, making contact only during brief windows of visibility.

On July 23, 1962, Telstar hosted the first live transatlantic TV broadcast. Viewers across continents simultaneously saw the Statue of Liberty, Eiffel Tower, a baseball game, and President Kennedy speaking. This event marked the dawn of global media connectivity through satellite communication, sidestepping earlier technologies like ionospheric reflection and undersea cables.

Interestingly, Telstar's 2,600 MHz frequency became associated with long-distance communication, though the familiar AT&T dial tone wasn't actually derived from it–that's just a myth.

Telstar's promising career was cut short when the Starfish Prime nuclear test created a radiation belt that damaged its transistors. Engineers revived it temporarily by cycling power–like restarting a frozen computer–but by February 1963, Telstar fell silent. Its legacy, however, speaks volumes; it still orbits Earth today as a memorial to communication's future.

In 1963-64, Syncom satellites achieved what Arthur C. Clarke had predicted in 1945–geosynchronous and geostationary orbits. Like celestial parking spots, these positions allowed satellites to hover above fixed points on Earth. Syncom 3 demonstrated this power by broadcasting the 1964 Tokyo Olympics live to America, ushering in an era of continuous global communication.

The world had found its voice in space, and it was only getting louder.

Weather Watching (1970s)

Weather satellites began in 1960 with NASA's TIROS program that had cameras onboard to capture basic cloud images from space. These simple satellites revolutionized forecasting by giving meteorologists their first aerial view of Earth's weather.

In 1970, NOAA evolved the experimental TIROS program into a sophisticated weather monitoring network that featured satellites that carried infrared sensors (like thermal goggles seeing through darkness), radiometers (moisture detectors), and advanced cameras. They became Earth's watchful guardians, monitoring our planet's weather system like doctors checking vital signs.

These technological advancements saved countless lives, particularly with hurricane tracking. During Hurricanes Agnes (1972) and Eloise (1975), satellite warnings gave communities precious preparation time. Before this technology, storm warnings came with mere hours' notice—now people had days, like having a flashlight to navigate through the dark instead of stumbling blindly.

Weather forecasting changed forever. Previously, meteorologists worked like detectives with scattered clues from ground stations, ships, and airplanes. With satellites, they gained continuous global coverage—tracking storms from birth to death and watching weather systems journey across oceans long before landfall. This comprehensive view marked the birth of modern meteorology as we know it today.

Before GPS: The Carpenter's Clock

The story of satellite navigation begins with an age-old challenge—determining one's position on Earth. In the eighteenth century, John Harrison solved the longitude problem at sea with his revolutionary marine chronometer. This breakthrough made accurate timekeeping the key to geographical precision and later formed the foundation for satellite navigation.

Before GPS, navigation's greatest danger wasn't storms but position uncertainty. Sailors could determine latitude using the North Star, but longitude remained elusive and often led to disaster.

Earth rotates fifteen degrees hourly, so knowing the time difference between home and local noon theoretically allows longitude calculation. However, at sea, mechanical clocks failed—ship movement disrupted pendulums, salt corroded components, and temperature fluctuations warped precision parts. Without reliable timekeeping, navigators relied on dangerous guesswork.

After losing four warships near the Isles of Scilly in 1714, the British Parliament offered £20,000 to solve the longitude problem. While experts expected an astronomical solution, John Harrison (1693-1776), a Yorkshire carpenter with mechanical aptitude, believed a reliable clock was the answer.

Harrison spent decades creating four increasingly precise timepieces. His final version, H4 (1759), resembled a large pocket watch but incorporated revolutionary features for temperature compensation and shock resistance. On its Jamaica test voyage in 1761, H4 lost only five seconds over eighty-one days—good enough to determine longitude within one mile.

Despite this success, Harrison faced resistance from astronomers on the Longitude Board. Only after King George III personally tested Harrison's work did Parliament award him £8,750 in 1773.

Harrison's chronometer revolutionized navigation and global trade. Ships could now confidently follow efficient routes, reducing insurance costs and disasters while making sea lanes predictable commercial highways. The East India Company completed faster trips to Asia, naval fleets coordinated at sea, and the Royal Navy gained maritime dominance partly through these accurate timepieces.

Scientific exploration advanced dramatically. Captain Cook used a Harrison-style chronometer for remarkably accurate mapping, while Darwin's HMS Beagle voyage relied on twenty-two chronometers to record specimen collection locations. These clocks helped scientists map coastlines, chart ocean currents, study Earth's magnetic field, and create the first global climate datasets—transforming geography from approximation into exact science.

Flying Clocks: The History of GPS

Harrison's principle that time equals position remains fundamental to modern navigation. GPS satellites carry atomic clocks, and your phone finds its position by measuring time differences between satellite signals. Basically, GPS is Harrison's idea updated with modern physics and launched into space. "Time is the thing," Harrison once said. And he was right.

The story of GPS begins in 1957 with an unexpected breakthrough. After the Soviet Union launched Sputnik, Johns Hopkins University scientists William Guier and George Weiffenbach discovered they could track the satellite by analyzing changes in its radio signal frequency—a phenomenon known as the Doppler Effect, similar to how a train whistle's pitch changes as it passes by. This also suggested that satellites could not only be tracked from Earth but could also be used to determine positions on Earth.

This then led to the development of TRANSIT in 1960, the first satellite navigation system. Used primarily by the US Navy for submarine positioning, TRANSIT proved the concept's viability though it provided only intermittent coverage and required long observation times.

The real advancement came when Dr. Ivan Getting proposed using satellite time signals for navigation. This concept evolved into the Global Positioning System (GPS), which began as a military project rather than the consumer tool we know today. The first GPS satellite launched in 1978, but the system only reached full operational capability in 1995, initially reserved for military use. In 2000, President Clinton authorized civilian access to high-precision signals, transforming navigation forever.

Today's GPS consists of over 30 satellites in medium Earth orbit, traveling at approximately 14,000 km/h. Unlike what many assume, GPS is fundamentally a time-based system, not a tracking system.

How GPS Actually Works

At its core, GPS functions through incredibly accurate timekeeping and geometric calculations. Each satellite carries multiple atomic clocks using cesium and rubidium atoms that help define the very second on Earth. These satellites continuously broadcast two critical pieces of information: their exact position in space and the time the signal was transmitted.

Your GPS receiver determines your position by measuring how long signals take to arrive from multiple satellites. It multiplies each signal's travel time by the speed of light to calculate the distance to each satellite. This creates a series of spheres centered on each satellite—your position is at their intersection.

To determine latitude and longitude (a 2D position), your receiver needs signals from at least three satellites. For altitude information (a 3D fix), a fourth satellite is required. This additional measurement also helps correct timing errors between your receiver's relatively inexpensive clock and the satellites' atomic clocks, eliminating the need for an expensive atomic clock in your device.

Interestingly, GPS must account for Einstein's theory of relativity. Because satellites move at high speeds far from Earth's gravitational pull, time flows differently for them than for us on the ground. Without adjusting for this effect, GPS measurements would quickly become inaccurate by kilometers.

Modern GPS accuracy improves through sophisticated techniques like differential GPS, which uses ground stations at known locations to correct satellite signals. This complex processing delivers precision within meters under good conditions.

Today's receivers further enhance accuracy by combining signals from multiple navigation systems beyond GPS. These multi-constellation devices employ advanced signal processing to maintain accuracy even in challenging environments like urban canyons or dense forests, making space-based timing signals precise location data that powers countless applications in our daily lives.

Even with its incredible capabilities, GPS has some limitations. Signals traveling from space to Earth encounter various obstacles that can impact their accuracy. Still, it's an amazing tool we rely on daily!

In urban environments, tall buildings create signal reflections (multipath effects) that confuse receivers. Earth's atmosphere bends signals as they pass through different air layers, causing distortions. Dense forests weaken signals, while mountains and tunnels block them completely. Device hardware also imposes limitations.

Under ideal conditions with clear skies, consumer GPS typically achieves five- to ten-meter accuracy. In challenging environments, errors increase significantly—explaining why your map location might appear inside a building while you're walking outside. The system itself isn't faulty; it's simply contending with real-world physical constraints.

From early signal tracking to today's satellite network, GPS has evolved beyond its creators' vision. Originally military technology, it now forms the foundation of our interconnected world. These orbiting satellites broadcast time signals democratically to anyone with a receiver.

This universal access to positioning has changed the way we navigate, live, work, and connect with each other. Every time you check a map, track a delivery, or share your location, you're engaging with a global system that turns timing signals into smooth, coordinated movement for billions of people—an unseen but vital thread binding modern life together.

The Global Impact of GPS

GPS has become society's invisible backbone—a digital nervous system spanning the globe. It synchronizes telecommunications networks, enables delivery scheduling, guides farming equipment, helps emergency vehicles find optimal routes, and ensures safe aircraft landings, influencing nearly every aspect of daily life.

The real-world impact is profound: farmers plant with pinpoint accuracy, search teams quickly locate disaster victims, and ships dock safely using automated systems. Every package delivery and food order relies on this technology working silently behind the scenes.

What began as an American innovation has evolved into a global endeavor. Multiple nations now operate satellite navigation systems: Russia's GLONASS provided Soviet-era independent navigation, Europe's Galileo offers civilian positioning, and China's BeiDou (completed in 2020) provides worldwide coverage.

Though each system operates independently with unique features, they function together seamlessly. Modern devices often use multiple systems simultaneously, combining signals for better accuracy in environments from dense cities to remote wilderness.

Together, these satellite constellations form the Global Navigation Satellite Systems (GNSS), pinpointing objects with centimeter-level precision anywhere on Earth—an extraordinary achievement based on synchronized clocks orbiting 20,000 kilometers above at 14,000 km/h.

While standard civilian GPS offers accuracy within meters, augmentation technologies like Real-Time Kinematic (RTK) and differential GPS achieve centimeter-level precision, enabling new applications from precision agriculture to autonomous navigation.

COSPAS-SARSAT: A Global Emergency Rescue System

Beyond navigation, a specialized satellite network focuses solely on saving lives–COSPAS-SARSAT. Created in 1982, this international emergency system detects and locates distress signals from emergency beacons activated by people in trouble.

COSPAS-SARSAT actually originated as a Cold War collaboration. Former adversaries–the Soviet Union and United States–along with Canada and France, united to establish this global emergency response network. Its name combines Russian "COSPAS" (*Cosmicheskaya Sistema Poiska Avariynikh Sudov*, meaning "Space System for the Search of Vessels in Distress") and English "SARSAT" (Search and Rescue Satellite Aided Tracking)–demonstrating how space technology can transcend politics for humanity's benefit.

The system works when someone activates an emergency beacon–whether lost in wilderness, stranded at sea, or experiencing aviation emergencies–satellites detect the signal, calculate its location, and relay information to ground stations that alert local rescue teams.

COSPAS-SARSAT has evolved significantly, now using three satellite orbit types: Low Earth Orbit (LEO) satellites for rapid detection, Geostationary Earth Orbit (GEO) satellites hover over the same region for constant regional coverage, and Medium Earth Orbit (MEO) satellites for fast, highly accurate positioning of the distress signal. This combined system can locate an emergency beacons within minutes, often narrowing the search area to around one-hundred-meters when GPS in included.

The impact has been substantial–COSPAS-SARSAT has helped rescue over 50,000 people worldwide. Whether hiking in Patagonia, sailing remote seas, or flying over the Yukon, emergency beacons connect to an international satellite network and rescue teams. This free, globally available service embodies the principle that when lives are at stake, politics and borders must not impede assistance.

Earth Observation Boom (1990s–2000s)

The 1990s and 2000s ushered in a new era for Earth observation satellite technology. During this time, the launch of increasingly sophisticated satellites dramatically advanced our understanding of the planet. A prime example is Landsat 7, launched in 1999, which marked a major leap forward in imaging capabilities. Equipped with its Enhanced Thematic Mapper Plus (ETM+) instrument, Landsat 7 could capture images with a resolution ranging from fifteen to sixty meters–good enough to identify individual buildings and monitor global changes in land use.

In 2002, the groundbreaking GRACE (Gravity Recovery and Climate Experiment) mission introduced an entirely new approach to studying Earth. Instead of relying on images, GRACE used two satellites flying in formation to measure minute variations in Earth's gravity field. This allowed scientists to track underground water movement, monitor ice sheet fluctuations, and even measure groundwater depletion in aquifers–insights unattainable through traditional observation techniques.

That same year, the European Space Agency launched Envisat, the largest civilian Earth observation satellite ever built at the time–about the size of a school bus. This colossal spacecraft carried ten different scientific instruments, effectively serving as a versatile space-based environmental laboratory.

Envisat's sensors enabled the study of ocean current temperatures, polar ice thickness, atmospheric chemistry, and the spread of air pollution. Over its ten-year mission, Envisat provided vital data that significantly advanced scientific understanding of climate change and environmental dynamics.

The Modern Era (2010s–Present)

The late 2010s marked the start of a revolution in how satellites are deployed, with the rise of satellite "mega constellations." Instead of relying on just a few expensive, large satellites, mega constellations use hundreds or even thousands of smaller, affordable satellites working together as a network. This approach is already changing life for people everywhere–for example, students in remote Alaskan villages can join online classes, and families in rural Scotland can stream movies, thanks to new global satellite internet coverage.

SpaceX led the way, launching its Starlink network in 2019. The first mission started with sixty small satellites, but thousands have been launched since then, making Starlink the biggest satellite constellation in history–containing over 7,000 satellites as of 2025, with plans for many more. These satellites orbit much closer to Earth than traditional ones, delivering faster internet with less lag.

Soon after, other companies joined in. OneWeb now has hundreds of satellites and plans for more, after overcoming early financial struggles. Amazon's Project Kuiper is building its own network and expects more than 3,000 satellites to go online in the next few years. Established companies like Telesat and SES are developing similar systems.

These modern satellite networks are highly advanced. Satellites now use powerful laser links to "talk" directly to each other in space, creating a fast, global "internet in the sky" that doesn't depend entirely on ground stations. They are also equipped with collision-avoidance systems that let them automatically adjust their paths–like self-driving cars–to avoid colliding with other satellites or floating space debris.

The reach of these networks goes far beyond providing fast internet. They connect schools in the countryside, deliver telemedicine to rural clinics, support rescue communications after natural disasters, and help farmers in developing countries check weather and markets or optimize their crops using satellite data. Satellites also help monitor environmental problems–like illegal deforestation–and give early warnings about disasters, protecting coastal and vulnerable communities worldwide.

However, this rapid expansion brings new challenges. Managing the huge number of satellites–much like air traffic control for space–has become a critical issue. New policies and systems are needed to keep orbits safe, prevent collisions, and address concerns like the growing risk of space debris. There's also an ongoing push to reduce how much satellites interfere with astronomy and to preserve clear views of the night sky for science and stargazing.

Overall, satellite mega constellations are helping bridge the digital divide, making societies more equal and resilient worldwide. Today's satellites support global cooperation and make it possible to respond quickly to climate issues, disasters, and humanitarian needs. What used to be science fiction is now everyday reality: the world is connected by thousands of satellites working together high above Earth, and this technology will only keep improving as time goes on.

Chapter 20

Anatomy of a Satellite

The Core Components That Power Our Space Technology

Modern satellites are marvels of engineering—machines built to thrive in the harshest environment, orbiting high above Earth's surface, bristling with technology and keeping us connected to the world below. Like miniature space stations, each satellite is packed with precision systems working in concert: atomic clocks that anchor global navigation, solar wings that chase the Sun for power, computers that think for themselves, and control systems that keep everything pointed just right. Let's peel back the layers and see how these orbiting wonders really work, starting with their incredible timekeeping system.

The Heart of Satellite Navigation

At the turn of the twentieth century, Max Planck found that energy was not an unbroken stream, but actually arrived in discrete, countable "quanta"—like coins dropped one by one into a slot. This meant that atoms don't vibrate at random but jump between precise, quantized energy levels, each jump giving off or absorbing energy at an exact, measurable frequency. Atomic clocks harness this quantum certainty. They measure time by tracking the ultrafast "ticking"—billions of perfectly regular oscillations every second—inside atoms like cesium-133 or rubidium-87, each oscillation driven by nature's most fundamental rules of quantum mechanics.

The first atomic clocks appeared in laboratories in 1949, built around ammonia molecules. It was Louis Essen (1908-1997) in 1955 who created the first practical cesium atomic clock–a device accurate to one second in over 300 years. By 1967, atomic clocks had become so reliable that the world redefined the meaning of a second: now, it is the time it takes for a cesium atom to vibrate exactly 9,192,631,770 times.

At the heart of every satellite is an atomic clock. Every GPS satellite broadcasts the exact time from its onboard atomic clock. Your device–phone, car, or ship–measures how long it takes those signals to arrive from different satellites, each signal racing at the speed of light. A timing slip as tiny as a billionth of a second will throw your location off by thirty centimeters. Take atomic time away, and the entire architecture of modern navigation–along with air travel, shipping, precision farming, and emergency response–just wouldn't work.

But that's not all folks: atomic clocks synchronize the global power grid, keep the internet humming, and provide ultra-accurate timestamps for financial transactions that power the world's markets. This relentless, hidden heartbeat in the sky is now so woven into daily life that its influence touches everything from guiding aircraft, to enabling international banking, to setting the pace for science and industry everywhere precision matters. Everything that hinges on exact timekeeping, in space or on Earth, relies on the quantum leaps first uncovered by Max Planck and carried aboard today's satellites.

Of course, perfect timing means little without the power to keep it running.

Keeping Satellites Powered in Space

Satellites need a stable, reliable power supply for their instruments and systems, but in space's vacuum, there are no power outlets. Their main energy source is solar panels: large, wing-like structures covered in photovoltaic cells that convert the Sun's rays directly into electricity. These panels pivot and track the Sun as the satellite orbits Earth, absorbing as much energy as possible to keep instruments operational and data flowing home.

The era of solar-powered satellites began with Vanguard 1 in 1958. About the size of a grapefruit, it was the first satellite equipped with solar cells and operated for over six years—demonstrating that sunlight alone could sustain a spacecraft far above the clouds. Vanguard's success led to increasingly resilient and sophisticated solar power solutions for space.

Solar panels had to be toughened to withstand space's harsh conditions: intense radiation, extreme temperature fluctuations, and impacts from debris. Engineers developed "maximum power point tracking," —something that sounds like corporate overlord governance—allowing the system to adjust each cell's output dynamically and maximize energy extraction from every photon.

The most significant advancement came with multi-junction solar cells. Unlike older single-layer cells that captured only one narrow band of sunlight, modern multi-junction cells layer several semiconductor materials. Each layer's bandgap is tuned to a specific part of the spectrum—blue and ultraviolet at the top, visible in the middle, infrared below—enabling more of the Sun's energy to be converted into usable power. Made from combinations of indium gallium phosphide, indium gallium arsenide, and germanium, these advanced cells exceed 30 percent efficiency—leaving older silicon panels at around twenty percent—while also reducing weight and saving space.

When a satellite enters Earth's shadow and loses sunlight, its backup power comes from onboard batteries—most often, the same lithium-ion type found in smartphones or electric cars, but designed to endure space's rigors. These batteries store excess solar energy during daylight and provide power during the dark, cold periods behind Earth, ensuring instruments remain operational and the satellite can function continuously.

The materials and designs developed for the demanding environment of space now support rooftop solar panels, solar farms, and energy-dense batteries in electric vehicles and portable devices. The same chemistry that helps satellites survive months without ground contact also allows us to stream music or travel cross-country without power concerns.

On Earth, the solar and battery technologies engineered for space's extreme conditions have been adapted for use on rooftops, in solar farms, and in electric and portable devices. Today, from remote villages using solar lanterns to advanced battery storage in electric vehicles, these innovations from space are helping fuel the transition to cleaner, more resilient energy for everyone.

Managing Heat in the Vacuum of Space

Space presents unique thermal challenges. Unlike Earth, where air distributes heat, space is a vacuum—heat moves only through radiation. One side of a satellite facing the Sun can reach 120°C (248°F), while the opposite side plunges to -100°C (-148°F).

Without air, heat can't circulate. To survive this, engineers wrapped satellites in "space blankets," gold-coated Multi-Layer Insulation (MLI) made from Mylar and Kapton, to manage these extremes. Developed in the 1960s, these blankets reflect sunlight while maintaining stable internal temperatures. We'll explore this further in our Space-Age Materials section.

Heat pipes arrived in the 1970s—devices that move heat from hot to cold areas using evaporation and condensation of fluids, without moving parts. Radiator panels emit heat directly into space. Advanced radiators feature adjustable louvers that control heat release.

Thermal control keeps evolving. Modern satellites incorporate phase-change materials that absorb or release heat as they melt or solidify, stabilizing temperatures during eclipses or engine burns. Transpiration cooling mimics sweating—pushing fluid through microscopic holes on the spacecraft's surface, cooling as the fluid evaporates into space.

These cooling technologies found uses on Earth across computing and medicine. Supercomputers rely on heat pipes. Vaccine storage facilities use phase-change materials. Computer chip manufacturers adopted space-inspired cooling systems.

The Brains of Satellites

Every satellite depends on a computer built for the unforgiving conditions of space—a world where radiation routinely zaps ordinary electronics. The earliest satellites in the 1950s barely needed brains: Sputnik didn't have a computer at all, just enough circuitry to beep.

But as missions grew more complex, so did their need for real-time thinking.

The Apollo Guidance Computer (AGC) in the 1960s was a compact, robust system built by MIT and NASA to help astronauts find their way to the Moon and back. It proved once and for all that computers could withstand radiation storms and power surges, making split-second decisions when Earth was hundreds of thousands of miles away.

Today's satellite computers are in another league entirely. Packed with self-checking and fault-tolerant design, these radiation-hardened modules thrive where off-the-shelf chips would quickly fail. They run redundant calculations in parallel, spot errors before they cause trouble, and can detect, diagnose, and even repair glitches without waiting for help from mission control. These onboard "brains" orchestrate everything: steering solar panels to catch the Sun, keeping instruments at just the right temperature, sending scientific gold home to Earth, and making fast, independent decisions about power, pointing, and propulsion.

Over decades, this survival-driven tech has re-entered our daily lives in amazing ways. The safeguards that keep a Mars probe alive now enable pacemakers to beat reliably, self-driving cars to steer, and robots to operate safely in factories. When your washing machine balances a spin cycle or your phone plots a cross-country route, you benefit from technology designed to keep satellites working perfectly in space.

Staying on Course

Accuracy in orbit is really important. Each satellite needs to stay in its spot and keep its direction with incredible precision, all while zipping through space faster than a speeding bullet. Because even a tiny misalignment can lead to missed images, lost signals, or incorrect data. Whether they're taking pictures, transmitting calls, or monitoring hurricanes, satellites have to stay perfectly aimed at all times.

Modern satellites use *attitude control*, which is basically a high-tech autopilot, to keep themselves aimed in exactly the right direction.

Inside each spacecraft, reaction wheels spin up or down, shifting the satellite's orientation with astonishing accuracy—even without burning fuel. Gyroscopes continuously sense twists and turns, helping the system maintain stability, much like a spinning bicycle wheel steadies a bike. We'll explore these concepts in Spin Class later in the book.

For more muscle, satellites use magnetorquers—devices that create magnetic fields, letting the spacecraft push against Earth's own magnetism. When bigger changes are needed, fast bursts from tiny thrusters can be used, pivoting the entire satellite with brief jets of propellant.

This hardware is a far cry from the early days, when satellites like Explorer and Vanguard relied on simple spinning to keep steady. Today's computer-guided systems can maintain pinpoint accuracy through years of orbit, counteracting everything the space environment throws at them.

But orientation is only half the problem. Satellites must also keep to their tracks in space—*station-keeping*—despite forces trying to knock them off course: atmospheric drag in low orbits, sunshine pushing in higher orbits, gravitational tugs from the Moon and Sun. Two of our rocket engine friends help satellites do this: classic chemical thrusters for quick, powerful pushes, and ion engines that offer a more efficient, sustained, and gentle touch.

This allows new satellite constellations such as Starlink to orchestrate the thoughts of spacecraft at once, each one individually guided and updated, each avoiding space debris and delivering reliable global coverage.

Satellites are more than machines orbiting Earth—they are timekeepers, power plants, thermostats, computers, and navigators all in one. Every part reflects a piece of human ingenuity, working together to keep a fragile system alive in the harshest environment imaginable.

Satellites embody our greatest scientific truth: when we learn to survive in space, we learn how to thrive on Earth.

Chapter 21

Beyond Vision

How Satellites See Our World

High above Earth, observation satellites act like our planet's extended senses–forming a worldwide network of artificial eyes, ears, and nerves that keep us updated about what's happening everywhere. Whether it's wildfires blazing in forests or tiny chemicals floating invisibly in the air, these satellites gather vital information. Each one has its own special way of helping out–some capture familiar images with cameras, others sense heat or radio signals, and some "sniff" the air to detect trace gases. All together, they translate a constant stream of signals–images, pulses, patterns–into early warnings, helpful guidance, and valuable insights for those of us on the ground.

A key to their power is how they "see." Passive satellites, like cameras, detect light or heat that Earth reflects or emits naturally, revealing the planet much as our own eyes do. Active satellites go further, sending out pulses of energy–whether laser, radar, or microwave– and measuring what bounces back. This dual approach allows the satellite fleet to work around the clock: passive sensors capture the world in daylight, while active sensors cut through darkness, storms, and smoke, probing places ground-based systems cannot reach.

These orbiting sentinels reveal almost everything: wildfires as they ignite, crops greening across continents, ice sheets cracking, or methane leaks escaping pipelines. Their insights go beyond curiosity or defense–modern satellites are now central to disaster response, environmental policy, resource protection, and public safety.

Each sensor type offers a unique superpower. Optical imagers provide a bird's-eye view, radar reveals the invisible, thermal sensors track heat and energy, spectrometers decode the chemical makeup of air and land, and quantum or gravimetric instruments uncover hidden forces shaping our world. By combining these senses, satellites create the first continuous, dynamic, and three-dimensional picture of Earth—one that no ground-based system could ever match.

Taking Pictures

Satellites peer down on Earth with some of the most advanced optical cameras ever made, providing a perspective no airplane or human eye could match. From orbits hundreds of kilometers overhead, their cameras capture detailed, color-rich images—showing forests, oceans, fields, roads, and city lights with an astonishing clarity. These images let us monitor the spread of wildfires in real time, measure the growth of cities, track how crops are faring across an entire continent, and even map the aftermath of natural disasters, all from the vantage of space.

Optical satellite imaging started in 1960 during the Cold War, with the CORONA program using film cameras on spy satellites, tossing return capsules through the atmosphere to be recovered mid-air. In 1972, the focus changed when NASA and the USGS launched Landsat 1—not for espionage, but to support scientific research. This public mission was a game-changer, transforming Earth observation from a military tool into a valuable resource for science and society, especially after infrared sensing was incorporated.

Prominent figures such as Dr. Virginia Norwood, who created the Landsat Multispectral Scanner System allowing satellites to capture both visible and infrared images, and Dr. William Pecora, a strong advocate for making satellite data freely available, have really revolutionized environmental research. Their commitment to openly sharing satellite images has provided us with amazing new tools to help firefighters battling wildfires, farmers coping with droughts, and scientists uncovering vital clues about climate change.

Today's optical imaging satellites are marvels of miniaturization and power. Commercial satellites like WorldView-3 can pick out objects as small as thirty-one centimeters, allowing us to see a parked car or the health of a field with astonishing clarity. Europe's Sentinel-2 satellites do more than just take pictures—they capture thirteen different bands of light, from visible reds, greens, and blues to near-infrared and shortwave-infrared. This helps detect everything from toxic algae blooms to hidden pests in crops. These satellites are incredibly useful—they track wildfires as they start, keep an eye on floods as they happen, uncover illegal mining activities, help farmers assess water and soil quality, and create detailed environmental maps that support urban planning and conservation efforts.

Modern satellites are equipped with onboard computers that process imagery as they are collected, quickly detecting unusual changes and notifying disaster response teams and scientists well before humans on the ground even become aware of what has happened. From spy film capsules to these smart, real-time observers, optical satellites now act as Earth's vigilant eyes in the sky—transforming raw images into information that helps us protect our rapidly changing world.

Hidden World

While traditional satellite cameras display Earth in familiar shades of red, green, and blue, hyperspectral imaging satellites go beyond what is normally visible. These imaging sensors split sunlight into hundreds of closely spaced spectral bands, allowing scientists to identify the chemical composition of soils, plants, water, and even atmospheric gases. With this kind of data, it's possible to discover hidden mineral seams beneath deserts, monitor crop stress before leaves visibly wilt, or even map oil spills spreading across the ocean.

The roots of this technology go back to Sir Isaac Newton, who, in the seventeenth century, discovered that white light is made up of many colors. During the mid-twentieth century, spectral imaging developed quickly as military researchers recognized that each material leaves a unique "spectral signature" that can be used to identify camouflaged tanks, distinguish soil types, or detect otherwise hidden changes.

A major milestone was achieved when NASA's Earth Observing-1 satellite was launched in 2000, equipped with the Hyperion instrument which can distinguish over 220 individual bands of color and heat. Since then, commercial companies have made great progress with this technology. Planet's Tanager satellites help monitor greenhouse gases from industry, while Pixxel's constellation offers early alerts to farmers about crop diseases before they become widespread. A new wave of custom sensors is now emerging to assist with a variety of tasks, from insurance to resource exploration.

Modern hyperspectral satellites produce huge volumes of data, often referred to as "data cubes"—detailed location grids filled with hundreds of color-coded measurements at every pixel. These advanced satellites are equipped with not just cameras but also sophisticated onboard AI, enabling them to spot early signs of problems like toxic algae blooms, methane leaks, or forest fires from space, sometimes just minutes after they begin.

What started in laboratories and Cold War intelligence has now blossomed into a global toolkit dedicated to caring for our environment. Archaeologists can reveal faint outlines of lost cities hidden beneath jungle canopies. Climate scientists keep a close eye on how much carbon forests are soaking up. Mining companies can map out resources safely without stepping foot on the sites. Farmers get early alerts to help them respond when pests or droughts might threaten just one field.

In just a few decades, hyperspectral imaging has evolved from a purely scientific pursuit to an essential tool for capturing even the tiniest yet most important changes across Earth's diverse and lively environments, from individual plots to entire continents.

Bat Signals

How does a bat see at night? Without using its eyes, the bat emits a high-pitched squeak. The sound bounces off walls, trees, or insects and returns as an echo. By hearing these echoes, the bat can create a mental map of its surroundings—catching food and avoiding obstacles, all in complete darkness.

Synthetic Aperture Radar (SAR) satellites operate in a remarkably similar way, but in space and on a much larger scale. Instead of sound, SAR satellites send out pulses of microwave energy. These radar waves travel down to Earth, reflect off forests, mountains, buildings, or even tiny ground shifts, and then return to the satellite's antenna. Just as a bat pieces together echoes from all directions, SAR collects these returning signals and constructs detailed images and measurements—even when thick clouds or darkness block the view of visible cameras.

While a bat emits hundreds of calls each second to gather more information, SAR satellites fire thousands of radar pulses from different angles. While the bat's brain turns each echo into a real-time "picture" of insects and trees, a SAR satellite's onboard computer combines all its echoes to "see" floods under heavy rain, map glaciers shifting millimeter by millimeter, or detect subtle bulges in volcanoes that indicate impending eruptions.

The roots of this technology go back to the Cold War, when military planners needed an eye in the sky to see targets hidden by weather or darkness. Carl Wiley first conceived of the idea in 1951, with Nobel laureate Luis Alvarez's team helping develop practical systems. NASA's SEASAT-A carried the first SAR into orbit in 1978, and spacecraft like Magellan and Cassini later used radar to penetrate the clouds of Venus and Titan, revealing landscapes that humans could not otherwise see.

What makes SAR unique is its ability to combine data from dozens, even thousands, of pulses at different angles—effectively simulating a giant radar dish stretching kilometers across, even though its actual antenna might only be a few meters wide. This trick provides satellites with the incredible sensitivity needed to detect changes the width of a paperclip from hundreds of kilometers in the sky.

Today, SAR has moved from top-secret to an important tool for monitoring disasters—from Bangladesh floods to polar ice shifts while a large share of Europe's Sentinel-1 SAR data is freely available. NASA and India's upcoming NISAR mission will combine two radar bands—L-band for deep penetration into trees, soil, and ice, and S-band for ultra-fine surface detail.

Private companies like Capella Space, ICEYE, and Umbra miniaturize SAR technology, deploying swarms of small satellites that can provide sub-meter detail anywhere on Earth, regardless of weather, sometimes updating images within minutes. AI now analyzes SAR data to detect changes and send instant alerts for floods, landslides, illegal logging, or ship movements.

What began as a simple spy tool has grown into one of our most vital resources for protecting our planet. SAR helps us predict floods, keep an eye on glaciers, alert us to earthquakes, guide first responders, and ensure the safety of global shipping. By uncovering the hidden aspects of the world—whether it's day, night, or stormy weather—SAR satellites serve as a worldwide early warning system, offering exciting new ways to safeguard all life on Earth.

Laser Maps

LiDAR (Light Detection and Ranging) is like a super-precise laser ruler floating around Earth. Instead of capturing images with cameras or using radio waves like radar, LiDAR satellites send out quick bursts of laser light aimed at the ground and measure exactly how long it takes for those pulses to bounce back. By firing hundreds of thousands of laser pulses every second, LiDAR creates detailed 3D maps of everything—city skyscrapers, mountain valleys, expansive forests, and winding rivers—with incredible accuracy, often within just centimeters.

The real power of LiDAR becomes evident when engineers can track not only the distance but also exactly where the sensor is and how it's moving. First developed in military labs during the 1960s, LiDAR became really useful when scientists combined laser scanners with satellite positioning (GPS) and onboard motion sensors like gyroscopes and accelerometers (also called inertial measurement units or IMUs). With GPS providing precise position and IMUs capturing tilt, turn, and acceleration, each laser pulse can be accurately placed in 3D space. This combination turns billions of simple distance readings into highly accurate, georeferenced maps of terrain, cities, and forests.

Today's orbital systems like NASA's ICESat-2 shoot 10,000 laser pulses per second to detect tiny changes in polar ice, tracking sea level rise with millimeter-level changes. Meanwhile, the GEDI system on the International Space Station scans forests, helping estimate carbon stored in trees as part of the fight against climate change.

LiDAR's incredible ability has uncovered many hidden treasures of the past: ancient Maya cities tucked away in the jungle, forgotten temples at Angkor, and even clues to vanished rivers beneath our cities' concrete and soil. Today, it's also a vital tool for assessing flood risks, monitoring the health of bridges, guiding self-driving cars, and helping identify areas prone to landslides or earthquakes.

As new LiDAR satellites are launched and onboard computers become even more advanced, this technology grows even more powerful. In the near future, space-based lasers combined with AI will be able to spot changes within minutes and warn officials about emergency situations as they happen. Today, you can see portable versions on drones and smartphones, but nothing beats the incredible reach and accuracy of the large LiDAR systems orbiting high above. They are creating the most detailed maps of our constantly changing planet, helping us understand it better every day.

Atmospheric Sensors

Earth observation satellites do so much more than just monitor ground activity; they also observe the atmosphere. Acting as orbiting chemistry labs, they decode our planet's atmospheric secrets. As sunlight passes through different air layers, each gas absorbs or bends the light in a unique way, leaving behind a kind of "spectral barcode" or fingerprint that satellites can read from space.

To reveal these atmospheric fingerprints, satellites use two main tools. Spectrometers split sunlight into a rainbow of colors, allowing scientists to detect faint, telltale shadows where gases like carbon dioxide or methane have absorbed specific wavelengths. Radiometers, on the other hand, measure the intensity of this light, tracking subtle temperature and energy shifts that influence Earth's climate and weather.

In the 1800s, pioneers like Fourier and Arrhenius understood that Earth's atmosphere traps heat—what we now call the *greenhouse effect*. Later, satellites enabled us to study these effects from space. One memorable moment came in the 1980s, when satellites uncovered a significant thinning of the ozone layer over Antarctica, soon dubbed the "ozone hole." Ozone acts as a natural sunscreen for the planet, absorbing harmful ultraviolet (UV) radiation from the Sun. However, this protective layer was being destroyed by chemicals called chlorofluorocarbons (CFCs), common in refrigerants and aerosol sprays. Satellite data also told us that the ozone hole was expanding each year, allowing more UV radiation to reach the surface, potentially harming living organisms and disrupting ecosystems. This urgent warning prompted international action: the 1987 Montreal Protocol, a global agreement to phase out CFCs and protect Earth's delicate ozone layer. It showed how a space-based discovery could unite nations and change policies to safeguard our planet's health.

Today, atmospheric satellites are more advanced than ever. NASA's OCO-2 tracks where carbon dioxide is produced or absorbed, CrIS creates 3D maps of temperature and moisture to model weather, and satellites like OMPS and Sentinel-5P monitor the ozone layer and scan for hazardous pollution. Their data help us forecast wildfires, detect volcanic ash clouds that threaten aviation safety, and can help identify toxic gas leaks before they spread.

The future is even more exciting: soon, satellites will increasingly use AI to read Earth's "air barcodes" in real time, alerting us instantly to pollution spikes or dangerous weather conditions. From Cold War experiments to vital public services, these space-based chemistry labs have become some of our most important tools for protecting health, safety, and the environment.

Quantum Ways

This is just going to be weird to understand. All things "quantum" are a bit mind-bending. Nevertheless, we need to talk about it as it does represent some of the latest and most innovative technologies ever invented.

What's happening in quantum sensors marks a major step forward in how we observe Earth from space. Instead of regular cameras, these sensors use atoms themselves–detecting changes too subtle for any other instrument. At their core, quantum sensors use the unique properties of quantum mechanics. Atoms can, in a sense, exist in more state or one place at once (a phenomenon called *superposition*), and can interact with each other to create interference patterns—revealing incredibly tiny details about their environment.

And they are so sensitive; it is like your friends could sense you're whispering from another city. They can detect minute changes in gravity and magnetic fields, the kind of changes a regular scale couldn't pick up, like adding a single grain of sand to a beach. For example, a quantum gravimeter cools atoms with lasers until they nearly stop moving, then releases them and tracks their fall. Even the slightest pull from a rock, an underground water pocket, or shifting magma under a volcano leaves a measurable mark on how those atoms move. This method is so accurate it can detect things like hidden aquifers, shifting tectonic plates, or magma flows that might signal an upcoming earthquake.

Quantum sensing builds on major physics breakthroughs—starting with Schrödinger's famous wave equation from 1926 and later advances in atomic clocks and laser cooling by Nobel-winning scientists like Steven Chu and Claude Cohen-Tannoudji. In the late twentieth century, researchers showed that interfering atomic waves could measure gravity with a sensitivity once thought impossible. By the 2000s, satellite missions such as ESA's GOCE measured tiny gravity variations from space with incredible sensitivity, paving the way for future quantum sensors to operate on satellites.

As of 2025, full quantum gravimeters are being tested on rockets and aircraft, while new missions from ESA (CARIOQA) and NASA are racing to demonstrate them in orbit. CubeSats are now being equipped with durable quantum sensors suitable for space, even though these experiments began in shielded labs. Once these technologies become operational, the benefits will be immense.

Quantum sensors will help us monitor underground water resources, track ocean currents, assess earthquake risk, and perhaps even guide spacecraft or cars without GPS. They open a whole new window into forces like gravity and magnetism—offering us a new way to map, predict, and protect our dynamic planet, seeing deeper than any eye or camera before.

The Ultimate Human Mirror

This ongoing, global surveillance reflects the best of humanity's innovative spirit, helping us to see and understand our planet in ways our ancestors could have only dreamed of. These satellites overhead do more than just observe; they listen, measure, and even "sense" the delicate, unseen ripples passing through land, air, and water. Each new generation of satellites improves our view and allows us to respond faster, revealing what was once hidden—fires, floods, pollution, drought, and even tiny shifts in gravity—making these threats clearer and more manageable.

What started as tools for science and security have become trusted partners in our effort to protect and nurture our planet. Earth observation satellites now create a vibrant, living picture—covering food security, emergency alerts, climate initiatives, and public health—all enriched by their invaluable insights. By transforming incredible amounts of data into practical guidance, these systems help us focus on what truly matters, understand what's at stake, and take early action that can change lives.

Most importantly, these spacecraft remind us how deeply connected we are to the fragile, ever-changing world that supports all life. As we find new ways to see and learn, we naturally become more aware of our responsibilities: to respond with kindness, to protect with purpose, and most of all, to care deeply for our precious, ever-evolving planet.

Chapter 22

Baby Satellites

The Rise of Mini-Satellites: A Lego-like Revolution

Space was once the domain of giants—massive satellites crafted by nations and billion-dollar agencies, each as big as a pickup truck or even larger. But in 1999, two professors—Bob Twiggs from Stanford and Jordi Puig-Suari from Cal Poly—had a vision to make satellite building accessible to students and small groups without big budgets. Their solution? Lego bricks (sort of).

Actually, their answer was the CubeSat—a tiny, standardized satellite "brick" measuring just ten centimeters on each side. It's small enough to hold in your hand, yet powerful and tough enough to handle serious scientific work in the challenging environment of space. Think of it like building with Legos: each "1U" unit acts as a modular block, which you can stack, connect, and customize—whether it's 2U, 3U, 6U, or larger—allowing you to create more advanced spacecraft piece by piece, mission after mission.

The era of CubeSats really began in 2003, when these miniature models first hitchhiked rides to orbit aboard Russian and European rockets. Initially, these tiny bricks were equipped with basic gear—a radio, a simple camera, maybe a sensor or two—serving more as proof-of-concept than a game-changer in the industry. Today, the landscape has shifted dramatically. Instead of a single agency or a global communications company controlling access to space, students, startups, and research labs can now design their own missions, mixing and matching components to match their needs and budgets. Suddenly, universities can launch weather stations, startups can test new communication technologies, and even high school students can participate in missions that were once only possible for superpowers.

During the 2010s, the Lego revolution really gained momentum. By 2012, engineers had successfully integrated powerful modern chips, advanced sensors, and durable radios into the standard cube, all while keeping each block under 1.33 kilograms (2.9 pounds). Now, CubeSats become convenient, snap-together science platforms that easily fit in a carry-on bag.

Scotland's AAC Clyde Space was one of the early pioneers in this field. They created innovative radiation-hardened "bricks" and sturdy solar panels, ensuring their cubes could withstand everything from the freezing polar nights at -120°C to the intense heat of +120°C under the blazing sun—allowing them to keep working reliably for years. By the middle of the decade, their satellites happily stayed operational in orbit for five to seven years, much longer than the early models that often failed or reentered the atmosphere after about a year.

Meanwhile, Planet Labs have changed the game. Instead of dedicating all their resources to making each Lego brick indestructible, they envisioned something new: what if you launched an entire swarm of blocks all at once, each capturing images of Earth as it spun beneath? Their "Dove" satellites—each with a twenty-nine-megapixel camera—collaborate seamlessly, like Lego pieces fitting together, to craft a complete, constantly updating digital mosaic of our planet at a three- to five-meter resolution.

The Lego approach gained popularity quickly. Kepler Communications assembled radio-focused CubeSats, that transmit real-time data at up to one hundred megabits per second—a significant leap in capability for something so small. ICEYE, headquartered in Finland, creatively incorporated Synthetic Aperture Radar (SAR) into its compact, brick-sized satellites, allowing their network to "see" through clouds and operate effectively at night. Meanwhile, Spire Global enhanced their satellites with weather-sensing modules, gradually building detailed atmospheric models by harnessing signals from faraway GPS satellites that are bent as they pass through Earth's atmosphere.

By 2020, over 1,200 CubeSat "bricks" spun in orbit. By 2023, that number doubled, as commercial rockets like SpaceX's Falcon 9 and Rocket Lab's Electron delivered hundreds of new Cubes to low Earth orbit each year—many assembled, customized, and programmed in classrooms and startup garages.

These modular satellites have already changed Earth science and disaster response: in Kenya, snapped-together satellites provide detailed crop data to farmers. Research teams monitor shifting glaciers brick by brick, and disaster responders rely on Cubes that can be swapped in to relay signals when ground towers go dark. Detecting wildfires, mapping emissions, rescuing communication—the Lego model adapts to every challenge.

The key to all this, of course, is accessibility. Building a CubeSat is (almost) as simple as assembling a Lego set: solar panels for power, radios for communication, straightforward orbital calculations, and creative tinkering for whatever mission is next. Add or remove blocks to fit your budget and imagination—a welcome change from the days when space was limited by price tags and engineering armies.

These snap-together satellites have turned space into a playground of invention. The old model—one big satellite for all—has been replaced by swarms of modular explorers, built and deployed at an ever-increasing rate.

Chapter 23

In-Orbit Servicing

Helping Space Sustainability

Throughout the Space Age, satellites were often seen as disposable. People would launch them, use their capabilities, and then set them adrift once their fuel ran out or systems failed, leading them to become space debris or burn up during reentry. However, a remarkable change is happening now. In-orbit servicing, repair, and manufacturing are opening up exciting new possibilities. These advancements let us keep, enhance, and even create new spacecraft right in space, rather than replacing entire satellites.

Extending a satellite's operational life is a simple and effective way to keep these valuable assets working longer. Usually, satellites don't fail suddenly; they just run out of the propellant needed to stay in the right position and orientation. Luckily, we now have specialized servicing spacecraft that can rendezvous with these inactive satellites, dock, refuel, or even take over propulsion! Northrop Grumman's Mission Extension Vehicle (MEV) has shown how well this works by docking with aging satellites and giving them several more years of service. Instead of becoming space debris, these satellites get a second chance, helping us maintain a sustainable and cleaner orbit.

Robotic servicing missions take this approach even further by doing repairs and upgrades right in orbit. They use advanced robotic arms that can replace solar panels, fix stuck antennas, or change instruments. These upgrades help satellites improve and adapt long after they've been launched. This flexibility creates a living, upgradeable space infrastructure—a flexible system that can respond to new needs and opportunities as they arise.

Orbital assembly is more than just servicing; it's about bringing together separately launched modules into big or intricate structures that can't be built on Earth. This opens up exciting possibilities: solar arrays sending energy back to Earth, large communications hubs, or deep-space exploration vehicles that are assembled in orbit before journeying to the Moon, Mars, or even further.

Many companies are pushing these innovations forward. Northrop Grumman is at the forefront with its MEV and upcoming servicing missions, while the Japanese startup Astroscale is dedicated to extending life and removing debris. SpaceLogistics LLC is working on fleet servicing solutions, and European leaders like Airbus and Thales Alenia Space are exploring robotic servicing and modular spacecraft technologies.

The growth of in-orbit servicing reflects a heartfelt commitment to space environmentalism. Refueling, repairing, and reusing satellites in orbit are wonderful ways to help slow down the rapid rise of space debris, which has become an increasingly urgent safety and sustainability concern. Companies like Astroscale are doing incredible work with debris capture missions and encouraging a circular economy in space—where satellites are refurbished, reused, or responsibly deorbited to keep our orbital environment healthy.

This exciting sector goes beyond just technological progress; it's opening up a vibrant new space economy. As startups and well-established aerospace companies work on robotics, fuel logistics, in-space manufacturing, and modular assembly, orbital infrastructure is rapidly becoming the backbone of global communications, navigation, Earth observation, and beyond. The capability to service assets directly in space not only brings economic benefits but also marks a significant stride toward environmental responsibility.

Of course, challenges still exist. Engineers are working hard to perfect docking techniques for satellites that weren't originally designed for servicing. Robotics must be resilient enough to operate in the tough conditions of space. Zero-gravity manufacturing calls for a fresh industrial approach. Standardization—like common docking interfaces and modular designs—will be essential, and it will take broad international collaboration to realize the full potential of on-orbit servicing.

As this new era unfolds, our connection with space is becoming more meaningful and lasting. Satellites are no longer seen as just throwaway tools; instead, they are valuable assets that we maintain and care for over many years. Just like how Earth's industries are now focusing on durability and repair rather than disposability, space is also embracing a more sustainable approach. This positive shift not only helps keep space activities viable for the long run but also safeguards Earth's orbital environment. Together, these efforts set the stage for a bright and sustainable future, inspiring generations of explorers to come.

Chapter 24

Working in Orbit

From Floating Labs to Future Habitats

Have you ever wondered what it's really like to live in space? It's not just floating weightlessly or taking stunning photos of Earth, though those are pretty memorable perks. Over decades, our way of living off-Earth has evolved step by step: initially, short trips, then longer missions, and today, the dream of making space a true second home.

When astronauts seem to float inside a spacecraft, they're actually experiencing "microgravity." In orbit, gravity remains strong—about 90 percent as powerful as on Earth's surface—but astronauts and their spacecraft are falling around Earth together. With nothing pushing them up from below, everything floats in free-fall, causing objects and people to drift through the cabin.

Microgravity means there's just a tiny bit of gravity—enough for things to stay nearby, but not enough to pull you down. This isn't "zero gravity," though you'll often hear that phrase. Life in this floating environment takes some getting used to!

Initially, human spaceflight focused on short trips—Vostok, Mercury, Gemini, Apollo. These missions made history, but staying in space for more than a few days was a new challenge. The next step was orbital stations like Skylab, Salyut, and Mir. These early outposts were like camping in space: cramped, tricky, but really important for learning how to live with microgravity. Over time, astronauts and engineers proved we could handle long-duration space stays.

Then came the International Space Station—the ISS. Can you imagine building a huge house from hundreds of parts, assembled in orbit by astronauts from different countries, in a place that's hard to reach? Since 2000, people have lived aboard the ISS every day, making it Earth's first permanent address in space. This giant floating neighborhood was put together piece by piece, teaching us how to construct and operate large structures where gravity is minimal.

Living in space has some surprises. In microgravity, you get a little taller as your spine stretches out. Fluids shift upward—hello, "puffy space face!" Even ordinary activities like eating, sleeping, or using the bathroom become little adventures. Every solution that scientists and astronauts invent helps us build safer, smarter habitats for the future.

The ISS is more than just a place to live—it's a laboratory for discovery. Scientists use it to explore everything from fire that burns without an "up," to how plants grow and adapt to weightlessness, to how our bodies work when freed from Earth's pull. Research covers biology, medicine, chemistry, and even agriculture—all helping us prepare for life farther from home. We will explore some of these topics in the next few chapters.

Every day on the ISS is a new chance to learn. The skills and knowledge gained here pave the way for creating new homes on the Moon, Mars, and maybe farther out. Humanity is moving from being visitors in space to becoming residents—a transformation as significant as settling new continents on Earth.

And that's pretty amazing, don't you think?

Early Steps Toward Space Living

Living in space might sound like a modern concept, but it actually has roots in the daring explorers of the past. In the 1800s, brave balloon pilots soared to incredible heights where the air was thinner and colder than what we're used to. Around the same time, submarine crews discovered how to survive deep underwater, where the pressure could crush anything not built for it. Both balloonists and submariners quickly understood the importance of sealed cabins—like protective bubbles—that could keep them safe in environments nature never designed for humans.

Inside these bubbles, conditions had to be just right. Pilots and crews needed to control air pressure so they could breathe normally, keep temperatures steady, and ensure there was enough oxygen. These challenges were very similar to those faced later in space. When space agencies started sending astronauts into orbit, engineers looked at how submarines kept their crews alive underwater and how balloonists stayed safe high above the clouds. Ideas like making sure every hatch was tightly sealed, recycling used air, and maintaining a stable atmosphere all influenced early spacecraft designs.

Initially, spacecraft were small and cramped, mainly designed to keep astronauts safe during brief missions. As technology improved, these survival pods expanded into much larger space stations like Skylab, Mir, and the International Space Station. Today, astronauts enjoy comfortable habitats where they can live and work for months at a time, complete with sleeping quarters, kitchens, and workspaces. What began as simple protection from tough environments has grown into an exciting journey to create truly livable homes in the vastness of space—making the dream of long-term space living a wonderful reality.

The Earth Shines Above Us

Everyone knows that looking directly at the Sun can hurt your eyes, but here on Earth, we're lucky that our atmosphere scatters and softens the sunlight. That's what gives us the blue sky and familiar daylight we enjoy every day. Out in space, though, there's no filter—sunlight is incredibly intense and bright, almost like a stage floodlight. It appears pure white because it contains all the colors of the spectrum at once. When a spacecraft moves behind Earth, that dazzling brightness quickly dims, and the cabin slips into a deep shadow.

In the late twentieth century, American author Frank White gathered stories from astronauts and noticed that when crews traveled far enough to see Earth's full curve, they often experienced a profound shift in perspective. Instead of focusing on borders and nations, they spoke more about unity, fragility, and our deep connection. They saw Earth's thin atmosphere, the curve of the horizon, and the continuous flow of oceans and weather—a delicate, interconnected system that sustains all life below. Many astronauts feel a stronger sense of responsibility, realizing they are part of something much bigger than any single country or tradition. White called this the *Overview Effect*.

A similar perspective appears in the stunning images of Earth taken from space and the Moon. The early photos from Mercury showed our blue planet rising from darkness, while Apollo's famous "Earthrise" and "Blue Marble" pictures captured the imagination of millions, showing Earth as a bright, lone sphere floating in the vast blackness of space.

In 2021, Dr. Sian "Leo" Proctor made history as the first African American commercial astronaut and the first woman to pilot a commercial spacecraft on SpaceX's Inspiration4 mission. She describes "being bathed in Earthlight"—the soft light from Earth that softly brightens spacecraft and the Moon.

This phenomenon has been appreciated for centuries. Leonardo da Vinci sketched the faint glow on the dark side of a crescent Moon, realizing it was sunlight reflected from Earth, not light coming directly from the Moon. Over time, this ghostly glow was called "earthshine" and "earthlight," both referring to sunlight that reflects from Earth and then from the Moon back to us. From the ground, you can see this effect best on the crescent Moon—look closely and notice the bright arc wrapped around a darker center, with the "dark" part still faintly visible. The illuminated part is directly lit by sunlight reflected from the Moon, while the rest glows softly with Earthlight—sunlight bouncing from Earth. Historically, this faint circle inspired phrases like "the auld moon in the new moon's arms."

By the early 1800s, astronomers used "earthlight" in textbooks to explain this partial illumination, helping us learn about Earth's reflectivity and cloud cover. Later, in science fiction and space writing, "Earthlight" gained new popularity. Arthur C. Clarke named his 1955 novel *Earthlight*, set between Earth and the Moon, and astronauts and writers began to use the term more widely to describe the soft, bluish glow Earth projects into space.

Physically, Earthlight is sunlight reflected back into space from our planet. When sunlight reaches Earth, it bounces off clouds, oceans, ice, land, and the atmosphere. Clouds reflect the most light, and the atmosphere scatters shorter wavelengths–giving Earthlight its soft white glow with a hint of blue.

This scattering, called Rayleigh scattering, involves very small molecules like nitrogen and oxygen in our atmosphere. They scatter blue and violet light more than red and orange, filling the sky with a soft blue during the day. The oceans enhance this effect–they absorb red wavelengths and reflect more blue and green, making Earth appear bright blue-white from space. During sunrise and sunset, sunlight passes through more atmosphere, scattering blue wavelengths and allowing warmer reds and oranges to create breathtaking sunsets we love to stare at with loved ones, perhaps with a glass in hand.

For astronauts, this reflected light becomes part of their daily experience. Even on Earth's night side, spacecraft often sit in a soft blue haze, with Earthlight providing enough glow for their eyes to adjust after sunset. Dr. Proctor's story reminds us that the light that makes space work possible comes from the planet that sent us there.

While the Overview Effect offers a majestic view of our planet as a whole, Earthlight catches sunlight and sends it back into space with its unique fingerprints. Indeed, while Earthlight unites us in our understanding of our own wee spot in the universe, the same physics also helps us understand the atmospheres and surfaces of our celestial neighbors–a theme that will return when we turn to other worlds later.

When astronauts look down at Earth or sit in a space "room with a view," lit from outside, it's a quiet, comforting reminder that even in space, the most reassuring light–Earthlight–comes from home.

Chapter 25

Our Orbital Home

Origins of the ISS

The International Space Station (ISS) is like a big, albeit cramped, house floating about 400 kilometers above our planet. It's the largest human-made structure ever placed in space, and it's more than just a lab and living space; it's a shining symbol of how countries can work together.

Back in the 1980s, the US and the Soviet Union were busy racing to build their own space stations—Space Station Freedom and Mir-2. But everything changed when the Soviet Union dissolved in 1991. Facing economic hurdles and rising costs, a daring idea took shape: what if these former rivals joined forces, sharing their skills and resources? That bold vision kickstarted a new era of space teamwork. Before long, space agencies across the globe saw the many benefits of collaboration.

What started as a partnership between the US and Russia blossomed into a wonderful alliance involving fifteen nations. Each nation shared its unique skills and technology to make this project a success. NASA provided important connecting modules, cutting-edge labs, and the station's robotic arm. Russia contributed vital propulsion and control systems, backed by years of expertise in long-term space missions. The European Space Agency brought us the Columbus laboratory, sleek life support systems, and the Automated Transfer Vehicle for smooth cargo deliveries. This collaborative effort really highlights the power of working together across borders to achieve extraordinary things.

Canada's Space Agency has earned a great reputation for its expertise in space robotics, creating the incredible Canadarm2 and Dextre, which are vital for assembling and maintaining the station. Japan's JAXA took pride in designing and building the Kibō laboratory, the station's largest single module, complete with an external facility for experiments exposed to space and its own robotic arm.

Many partner countries like Brazil, Belgium, Denmark, France, Germany, Italy, the Netherlands, Norway, Spain, Sweden, and Switzerland contributed by adding scientific instruments, specialized technologies, and components. Their combined efforts helped make the ISS financially feasible and transformed it into a shining example of post-Cold War cooperation, showing that former rivals can come together to push the boundaries of human space exploration.

Construction started 1998 with two key modules: Russia's Zarya, meaning "sunrise," which served as a control center, and the US-built Unity, the first connecting hub. What followed was truly one of humanity's most impressive engineering achievements. Over the next thirteen years, astronauts from around the world conducted more than 200 spacewalks, assembling the station piece by piece. Each spacewalk was a balanced maneuver in zero gravity, as astronauts skillfully moved massive parts—like giant Lego blocks—while rushing through space at incredible speeds.

Stretching 109 meters (356 feet) from end to end—roughly the width of a football field—the ISS provides astronauts with a living area comparable to a cozy six-bedroom house, offering 388 cubic meters of space to relax and move around. Its enormous solar arrays resemble giant wings, spanning an impressive area bigger than a basketball court—2,500 square meters or 27,000 square feet. What makes this feat incredible is how engineers from around the world meticulously designed every part to fit perfectly together in the challenging vacuum of space, all while the station orbits Earth every ninety minutes at an incredible speed of 7.8 kilometers per second.

Life on the ISS

Life aboard the International Space Station is a unique blend of cutting-edge technology and everyday challenges. Orbiting about 250 miles above Earth, this tube-shaped structure serves as both home and laboratory for its crew. Its living space, arranged like a series of interconnected train cars, creates an extraordinary environment where astronauts adapt to living without gravity.

The ISS usually hosts six crew members, each with a compact sleeping pod about the size of a phone booth. Astronauts have mastered the art of microgravity living: they sleep in tethered sleeping bags and keep their belongings secured with Velcro and straps to prevent items from floating away. Even routine tasks like eating or brushing teeth require careful planning and special equipment.

Maintaining this orbital outpost demands constant vigilance and creativity. The crew performs regular safety checks using tools designed to detect any air leaks—a critical task since even the smallest breach could be dangerous. Their responsibilities range from cleaning solar panels to monitoring the station's position relative to space debris. Unlike on Earth, there's no maintenance crew to call; astronauts must handle all repairs themselves, from basic upkeep to complex technical problems.

This hands-on approach has inspired new innovations. Over time, NASA developed a "fix-as-you-fly" philosophy for space repairs, including specialized tools and methods tailored for zero gravity. Engineers created special space-grade epoxy resins and quick-setting materials that astronauts could safely use in orbit. These materials were first tested on Space Shuttle missions before becoming standard repair options on the ISS.

Remarkably, these space-born innovations have influenced industries here on Earth. By 2005, aviation giants like Boeing and Airbus had adapted related rapid repair techniques, cutting aircraft maintenance downtime significantly. The automotive industry followed suit, with companies like Toyota incorporating similar methods into their manufacturing and repair processes. What began as solutions to space challenges has become a driving force for improving repair and maintenance across numerous industries on Earth.

Perhaps the most extraordinary result of tackling space-based challenges has been the creation of technologies that not only support life in orbit but also transform our everyday lives here on Earth.

Tasty Space Food

The story of space food preservation begins in the early 1960s when NASA faced a critical challenge: how to provide astronauts with nutritious, tasty food that could withstand the harsh conditions of space. Inevitably nutrition may have won out over taste!

To tackle this, NASA partnered with Whirlpool Corporation to develop advanced freeze-drying and dehydration methods. Unlike basic dehydration, these new techniques preserved not only the food itself but also its flavor and nutritional value. The team designed special packaging, creative storage solutions, and preservation processes tailored specifically to keep food fresh and appetizing in the vacuum and temperature extremes of space. This groundbreaking work, supported by experts in food technology, revolutionized how food is preserved—both in orbit and back on Earth.

Today, these same technologies have transformed food preservation worldwide where commercial freeze-drying facilities employ similar methods to extend the shelf life of everything from fruits and vegetables to fully prepared meals. These advances are especially valuable in developing regions where unreliable power supplies make traditional refrigeration difficult. By extending shelf life from days to years while preserving over 90 percent of the food's nutritional content, this technology significantly reduces food waste. Organizations such as the World Food Programme now use space-derived preservation techniques to deliver critical nutrition to disaster zones and food-insecure populations, making a lasting impact far beyond the space program.

Water Recycling and Purification

The International Space Station's water recycling system is a significant milestone in space technology. Back in the early 2000s, NASA engineers Layne Carter and Donald Carter led the development of the Water Recovery System (WRS), addressing a critical challenge: how to maintain a reliable water supply in space. This innovative system recovers an impressive 90 percent of all water onboard— from urine and sweat to moisture from breath—significantly reducing the need for costly water shipments from Earth.

In the unique microgravity environment of space, liquids behave very differently. Water doesn't flow down; instead, it forms floating blobs, sweat clings to the skin, and steam doesn't rise, making water conservation really important, but also challenging. The ISS's recycling system captures and processes every possible source of water, creating an efficient model of water management that has since been adapted for use on Earth–in disaster zones, refugee camps, and remote clinics.

Each year, the system processes around 6,000 liters of water using advanced techniques such as vapor compression distillation and multi-stage filtration. The result is water that is often cleaner than the tap water most people drink on Earth. Thankfully!

This space-born technology has had a profound impact back home. Companies like Water Security Corporation have adapted NASA-inspired designs into portable water purification units. These units proved invaluable during major disasters, including the 2004 Indian Ocean tsunami and Hurricane Katrina in 2005. Today, similar systems provide clean water to remote communities in over twenty-five countries, revolutionizing access to safe drinking water in challenging environments around the world.

Space Medicine

Long-duration spaceflight brings its own medical puzzles. How do you check a fever, mend an injury, or scan a broken bone in zero gravity?

Space medicine research aboard the ISS has sparked groundbreaking innovations that have changed healthcare on Earth. Back in the late 1980s, NASA was faced with how to accurately measure astronauts' temperatures in zero gravity without using traditional mercury thermometers, which posed a safety risk if broken in space.

In 1991, Dr. Jacob Fraden played a special role in advancing temperature measurement technology by creating an infrared ear thermometer designed for spaceflight. The device is based on a simple but powerful idea: all objects emit infrared energy that correlates with their temperature. Inside the device, a specialized sensor detects tiny amounts of infrared radiation from the eardrum and turns this into an accurate temperature reading. Because it doesn't need to touch the ear or use a glass bulb, a quick scan of the ear canal is enough to get a precise measurement.

This type of non-contact thermometer, originally developed for space, helped pave the way for wider use of infrared ear thermometers here on Earth. By the early 1990s, hospitals started adopting similar devices, and over the next ten years, they became common in clinics and airports, especially for fever screening during disease outbreaks. By 2020, infrared thermometers had become a well-known part of public health efforts worldwide.

A similar path of innovation emerged with portable X-ray technology. In the mid-1990s, NASA faced another challenge: how to diagnose bone loss or injuries in space without the bulky traditional X-ray machines. Dr. Richard Webber and his team at NASA's Johnson Space Center tackled this by reimagining X-ray technology. Instead of heavy film-based machines, they developed digital sensors that were lighter and more compact. Their breakthrough came in 1998 with the first portable X-ray system, weighing just twenty pounds compared to traditional systems that could weigh hundreds.

The team also solved power problems by designing efficient X-ray tubes that could run on battery power and developed software to enhance image quality while using lower radiation doses. These innovations soon moved from space missions to Earth, with commercial versions used in ambulances, remote clinics, and disaster zones by the early 2000s. Today, portable X-ray devices save countless lives in locations where traditional medical imaging isn't possible.

These space-born medical technologies have boosted healthcare accessibility worldwide. Remote villages now have advanced diagnostic tools, emergency responders can deliver better care on the spot, and doctors can make faster, more accurate diagnoses. This is a perfect example of how tackling the unique challenges of space exploration leads to breakthroughs that improve life for everyone on Earth.

ISS Legacy

The ISS is expected to be retired around 2030, but its legacy will endure far beyond that date. The knowledge and experience gained from building, operating, and living on the ISS have laid the foundation for the next generation of space habitats—whether future space stations orbiting Earth, sustainable bases on the Moon, or even human settlements on Mars. The challenges it helped us overcome in life support, long-duration missions, and international collaboration are critical stepping stones for humanity's expansion into deeper space.

More than just a scientific laboratory, the ISS has become humanity's shared home in space—a home built piece by piece by nations once divided, maintained by hands from every corner of the world. It stands as a powerful symbol of what humanity can achieve when nations come together.

It has demonstrated that people from different countries and cultures can collaborate successfully in one of the most demanding environments imaginable. This spirit of international cooperation, combined with groundbreaking technological advances, has not only advanced space exploration but also fostered a global sense of unity and shared purpose.

The ISS proved that humans could live, work, and thrive in space for extended periods, unlocking the door to a sustainable human presence beyond Earth. Its story is a testament to human ingenuity, resilience, and cooperation—showing that by working together, we can reach new frontiers that would be impossible to conquer alone. This enduring partnership will continue to inspire and guide space exploration for decades to come.

Chapter 26

Bodily Functions

— ◆ —

The Great Biological Recalibration

The human body was never designed to leave Earth—unless you are in the camp of people who believe we started on Mars and headed off to Earth for more exotic climes.

Every cell, joint, and organ evolved with one consistent environmental feature: gravity pulling down at 9.8 meters per second squared. Remove that, and things begin to drift—sometimes slowly, sometimes dramatically—but always in ways that reveal how deeply tied we are to the ground we left behind.

Spend a few weeks in orbit, and your body begins to adapt. Spend six months, and it starts to forget. That's the paradox of microgravity: disorienting, degenerative, but strangely useful.

In space, your body undergoes dramatic changes. Without gravity pulling blood downward, it flows upward throughout your body. This causes noticeable changes—your face becomes puffy while your legs grow thinner. Sinuses become congested, and vision can blur as pressure increases inside your skull. Your spine stretches out, temporarily making you taller. Without constant resistance, muscles begin to waste away. Perhaps most concerning is how quickly bones weaken as they release calcium—this bone deterioration happens at an alarming rate of one to two percent per month, far faster than how osteoporosis develops on Earth. This rapid loss would concern even the most experienced medical specialists.

The cardiovascular system also recalibrates. The heart, unburdened by vertical blood flow, reduces its workload. The balance organs in the inner ear lose their orientation. The immune system gets sluggish. Sleep patterns fragment. Cognitive function remains high—until stress, isolation, or disrupted circadian rhythms (your body's internal clock)—start taking their toll. And while some of this reverses upon return, some don't. Long missions change you, even after touchdown.

Orbit, it turns out, is a fast-forward button on degeneration. What takes years to unfold in aging, injury, or illness on Earth can happen in weeks aboard the ISS. That makes astronauts ideal data points for understanding muscle loss, bone density decline, cardiovascular drift, and immune suppression.

NASA's countermeasures have become blueprints for Earthbound medicine. Resistance exercise protocols developed for orbit are now used in physical therapy clinics and senior care facilities. Bone density monitoring systems, refined in space, have improved osteoporosis diagnosis and treatment. Even the psychological data from long-duration missions has shaped how we design shift schedules for healthcare workers and isolation protocols for submariners.

But that's what makes astronauts invaluable. Space may break the body, but it teaches us how to heal it better back home.

The Daily Mechanics of Floating Life

Living in zero gravity might sound like fun, but the reality is quite different. Everything we take for granted on Earth works differently in space, and astronauts have to relearn even basic daily activities.

Take sleep, for example. Without gravity telling your body which way is down, getting rest becomes a unique challenge. Astronauts can't simply lie in bed—instead, they use special sleeping bags attached to the walls of the space station that keep them from floating around while they sleep. Many astronauts struggle with their sleep patterns in space, experiencing vivid dreams and confusion. This has led to valuable research that helps people on Earth, particularly in developing better hospital beds and helping shift workers manage their sleep schedules.

Eating in space requires careful planning and special techniques. Something as simple as eating soup could become dangerous if it floats away in droplets. Even everyday foods like bread aren't allowed because the crumbs could float into sensitive equipment. Instead, astronauts eat specially prepared meals that are packaged to prevent spills. Each container must be secured, and astronauts use special utensils designed for zero gravity. The space-food packaging techniques have influenced some shelf-stable medical supplies.

Perhaps the most important daily activity in space is exercise. Astronauts must exercise for two hours every day to prevent their muscles and bones from weakening in zero gravity. They use specially designed equipment like treadmills with harnesses to hold them down, stationary bicycles without seats, and the Advanced Resistive Exercise Device (ARED). The ARED is particularly interesting—it simulates weightlifting in an environment where weights don't actually weigh anything, using vacuum cylinders and flywheels to create resistance. This rigorous exercise routine isn't actually about staying fit for appearance's sake—it's essential for maintaining the strength and bone density needed to return safely to Earth.

The Hidden World of Cells in Space

Beyond what astronauts experience, there's an invisible transformation happening deep within the cells. When we remove gravity from the equation, our cells reveal an amazing new side of human biology. Without Earth's constant pull, the basic building blocks of our bodies begin to act in ways we never see on Earth.

Inside the space station, our genes express themselves in new patterns, similar to how a musician might play a familiar song in a completely different style. Cell division and growth take unexpected turns. The healing process becomes unpredictable: sometimes wounds take much longer to heal than they would on Earth, while other times they heal surprisingly fast. Our immune system also acts differently in space, sometimes becoming overactive like an overeager security guard, and other times moving too slowly to respond effectively.

One of the most surprising discoveries? Cancer research. In space, cancer cells naturally form three-dimensional structures that closely mirror how tumors grow in the human body. This is something scientists have struggled to recreate in Earth-based laboratories, making the space station an invaluable place for cancer research.

These conditions have led to breakthrough discoveries in medical research. Scientists from NASA and ESA have learned new things about stem cells—the versatile cells that can develop into different cell types. They've made progress in understanding tissue regeneration, which could help people recover from injuries more effectively. They've also found better ways to grow protein crystals, which is crucial for developing new medicines.

The impact of this space-based research is already evident in modern medicine. Scientists have developed improved antiviral medications and created better versions of insulin. The ISS serves as an extraordinary laboratory where we can observe biology in its purest form, free from gravity's influence. Each experiment teaches us something new about how our bodies work—knowledge that would be impossible to gain on Earth.

Mental Impact

The body experiences changes in space—and so does the mind. When muscles weaken in weightlessness, it can also affect morale.

Living in space brings its own special mental challenges that astronauts have experienced since the very beginnings of space travel. From Yuri Gagarin's historic first journey in 1961 to the extended missions on the ISS today, we've discovered that the mental and emotional effects of spaceflight can be just as important as the physical ones.

Have you ever thought about spending half a year in a space that's smaller than most houses? The ISS, while about the size of a six-bedroom home, might feel even more confined because astronauts can't just step outside for some fresh air. There's a constant hum from the life support systems, artificial lighting that mimics sixteen sunrises and sunsets each day, and the familiar faces of crew members become your entire world. Research from NASA and

Roscosmos has shown that this environment can influence sleep patterns, mood, and cognitive performance.

The psychological effects tend to develop through several distinct phases. During the first two to three weeks, astronauts often encounter the "Overview Effect"–a profound shift in perspective when they see Earth from space. This initial burst of excitement typically leads into a period of adjustment lasting one to two months, as crew members get comfortable with their new routines and surroundings. By the three- to four-month point, many astronauts go through what is called the Third Quarter Phenomenon, a challenging time when feelings of fatigue and homesickness can be quite strong. In the final weeks before returning home, astronauts often experience a mix of pride, accomplishment, and anticipation as they prepare for their journey back.

NASA and other space agencies have developed support systems to help astronauts cope. Since the early days of the ISS in 2000, regular video calls with family have become a welcome and important feature. The agencies also send care packages on resupply missions, provide access to streaming entertainment and news from Earth, and arrange weekly sessions with ground-based psychologists.

This research has proven invaluable beyond space exploration. Mental health professionals have adapted these techniques to help people in similar isolated conditions. Antarctic research stations now use NASA's psychological screening tools. Submarine crews benefit from similar support systems. Even during the global pandemic of 2020, space psychology research helped develop better protocols for managing extended isolation.

One of the more interesting findings has been how studying space psychology has helped shape our understanding of team dynamics. The ISS has become a unique laboratory for studying how diverse international crews can work together effectively in high-stress environments. This research has influenced everything from emergency response team training to corporate leadership programs.

Looking toward future missions to Mars and beyond, these psychological lessons become even more important. A Mars mission could last at least three years, with communication delays of up to

20 minutes each way. What we have learned from ISS missions helps design better spacecraft interiors, crew selection processes, and support systems for these ambitious future journeys.

Importantly, this research has evolved how we approach mental health on Earth. The techniques developed for supporting astronauts have improved treatment for people with depression, anxiety, and PTSD. Hospitals have redesigned intensive care units based on lessons learned about reducing environmental stress in space. Even everyday workplace design has been influenced by our understanding of how confined spaces affect human psychology.

The journey to understanding the mind in space continues. As we look to survive space and establish permanent bases on the Moon and Mars, these psychological needs will be just as important as any technological advancement.

Every mission teaches us how to care for the body, calm the mind, and carry the best of ourselves wherever we go.

Chapter 27

Space Salads

The Next Frontier of Food Production: Growing Life in the Void

Space farming is one of humanity's most exciting adventures, not just for exploration but for our future survival beyond Earth. While rockets and advanced technology make space travel possible, nurturing human life in space depends on our ability to grow food. Plants are incredibly important—they provide essential nutrition, produce oxygen, recycle carbon dioxide, purify water, and even help keep astronauts emotionally balanced.

The journey of space farming began all the way back in 1982, when Soviet cosmonauts aboard the Salyut 7 space station successfully cultivated Arabidopsis, a small flowering plant. They managed to complete its entire life cycle in microgravity, from seed to seed. This proved that plants could adapt to the unique environment of space, opening up exciting possibilities for sustainable farming beyond our planet.

Growing in Space

Growing plants in space comes with unique challenges, all rooted in the absence of gravity. Without gravity, roots lose their sense of direction, spreading in many directions instead of growing down. Water acts differently too; it forms floating droplets that can suffocate plants if not handled properly. To address this, scientists have created special watering systems that use pressure differences and capillary action—the way liquids naturally move through tiny spaces—to give plants just the right amount of water and nutrients. This makes sure that plants stay healthy, getting enough water without the risk of overwatering or drying out.

On Earth, gravity naturally circulates air and water—warm air rises, cool air sinks, and water drains through soil. In orbit, that movement stops, so engineers install special fans to keep air flowing and regulate humidity. They also use synthetic growth media instead of soil to anchor roots and distribute water and nutrients evenly, because in weightlessness liquids tend to float and cling rather than soak downward as they do on Earth.

Lighting has also been a significant challenge. Early on, experiments depended on sunlight passing through spacecraft windows, but this wasn't always reliable or strong enough to support healthy plant growth. By the early 2000s, NASA developed special LED arrays that offer the perfect light wavelengths for photosynthesis while also being energy efficient. These smart LEDs can simulate day-night cycles and even tweak their light spectrum to suit the different growth stages of plants.

Space agriculture has made great strides since those initial experiments. In 2015, NASA's VEGGIE program achieved a major milestone when astronauts successfully grew and ate red romaine lettuce aboard the ISS. This showed that fresh food could be grown in space, opening the door for future possibilities. Building on this success, more advanced systems like the Advanced Plant Habitat have been developed. This system includes very detailed environmental controls and can keep track of over 180 different factors affecting plant growth, helping us understand how to grow plants better in space.

Today, astronauts grow a diverse array of crops such as fast-growing leafy greens like kale and cabbage, which provide essential nutrients in limited space. Small fruits like cherry tomatoes and peppers have proven successful, contributing both nutrition and psychological benefits. Root vegetables, especially radishes, are popular for their compact size and quick growth cycles. Flowering plants like zinnias are grown not just for research, but also to bring life and comfort to the sterile conditions of space stations, helping improve crew morale.

These farm systems are based on Controlled Environment Agriculture (CEA), which combines engineering and ecology to manage every detail–temperature, humidity, airflow, carbon dioxide, and light. Automated nutrient delivery systems constantly monitor what each plant needs and adjust feeding schedules to support healthy growth. Specialized growth chambers protect plants from cosmic radiation while keeping conditions just right. In this closed-loop system, almost nothing goes to waste–everything is turned back into life, creating a sustainable cycle that benefits us all.

Space-Inspired Solutions

Technologies developed for space farming are already transforming agriculture on Earth, with pioneering companies across different countries applying space-inspired solutions in innovative ways.

In Singapore, companies like Sky Greens and Sustenir Agriculture have built vertical farms using LED lighting and automated climate control inspired by space-based systems. Sky Greens' rotating tower design maximizes sunlight for stacked crops, enabling year-round production with ninety-five percent less water. Sustenir operates fully enclosed farms powered by renewable energy, maintaining climate controls–mirroring the efficiency principles pioneered in space agriculture.

In the UAE, Pure Harvest Smart Farms and Emirates Crop One have adapted aeroponics systems derived from space station technology. These deliver nutrients as a fine mist directly to roots, slashing water use by up to ninety-five percent–a critical advantage in desert climates. Pure Harvest integrates solar-powered climate control and sensor networks to produce fresh vegetables year-round, strengthening local food security.

In Scotland, Intelligent Growth Solutions leads in AI-driven vertical farming. Their systems use machine learning to monitor plant health, fine-tune lighting, and adjust air conditions in real time. This allows efficient, low-energy crop production even during the region's dark winters.

The Netherlands, long a hub of agricultural innovation, applies space-inspired hydroponics through companies like PlantLab and

GROWx. PlantLab's closed-loop systems recycle nearly all water and nutrients, while GROWx uses automated sensors to optimize growth for crops like tomatoes, herbs, and microgreens—boosting yield while minimizing resource use.

Beyond cities, these technologies are reshaping life in extreme environments. Antarctic research stations now grow vegetables year-round using ISS-derived systems. Refugee camps in Jordan and Kenya deploy portable solar hydroponics for food security. Even abandoned warehouses in urban "food deserts" have been reborn as indoor farms providing fresh produce to underserved communities.

Together, these companies and technologies demonstrate how innovations born from the challenges of growing food in space are reshaping agriculture worldwide—enabling us to grow more food with fewer resources and in places once thought unreachable.

Seeds of the Future

Space agriculture is transforming how we see food production, moving beyond just high yields to focus on using energy, water, and nutrients more wisely. This approach is so important for tackling global food security. Circular agriculture—where every output is reused—is becoming more popular, with support from groups like the World Food Programme.

As we create new ways to grow food for missions to Mars and the Moon, innovations like radiation-resistant crops, energy-efficient LED lighting, and systems that recycle resources in a closed loop will be essential. These advancements not only help us prepare for life beyond Earth but also offer strong, sustainable solutions for farming challenges here at home.

By overcoming the tough challenges of space farming, we're finding smarter, more efficient ways to provide nourishment everywhere—using less water, energy, and land, and growing healthier food. Space farming is more than just agriculture; it's an essential step toward making humanity a multi-planetary species and ensuring a sustainable future for everyone.

Chapter 28

Weird Science

Where Everything Works Differently

Space provides scientists with a truly unique laboratory unlike anything else on Earth. In orbit, particularly on the International Space Station, the experience of weightlessness, or microgravity, changes how molecules, cells, and materials behave. This weightless environment lets life and chemistry develop in surprisingly clearer and often more insightful ways. Without gravity pulling liquids or cells downward, substances can drift freely and grow without the distortions we see on the ground. This special setting enables researchers to design new experiments, ask questions that are impossible to explore on Earth, and discover pathways to better medicines, therapies, and technologies. What begins as curious "weird science" in space quickly transforms into practical, life-changing discoveries for everyone.

The journey of microgravity research started during early space missions like Mercury and Apollo, when astronauts and scientists first observed that fluids, cells, and chemical reactions behave very differently outside of Earth's atmosphere. Over the years, orbiting labs like Skylab and Mir set the stage for the International Space Station, launched in 2000–the first permanent laboratory in space. Over the past twenty years, research teams aboard the ISS have explored biotechnology, pharmaceuticals, stem cells, cancer studies, and so much more.

Proteins and Cells

Space has allowed us to glimpse some of some of life's best-kept secrets. One standout discovery is how we can grow nearly perfect protein crystals in space. Here on Earth, gravity often causes imperfections in crystal growth, making it harder for scientists to study intricate structures. But in microgravity, crystals can grow larger and more perfectly—kind of like snowflakes on a chilly winter morning—with incredible clarity. This helps us see detailed molecular features that are essential for understanding diseases and developing new medicines. These space-grown crystals have played a key role in improving drug design, especially for tough-to-treat illnesses. In a way, space acts like a powerful molecular 'microscope' that deepens our understanding of life at the tiniest level.

Space also offers the ability to observe living cells in their natural 3D shapes. On Earth, cells in labs flatten on culture dishes, but in microgravity they form spherical clusters that more closely mimic tumors in the body. This provides researchers with a more realistic model for testing drugs and understanding disease progression, thereby improving therapies in ways previously impossible. This provides researchers with a more realistic model to test drugs and understand disease progression, improving therapies in ways that are hard to achieve on Earth alone. Extensive studies on how microgravity affects astronauts' bone density and muscle mass have also contributed to better strategies for treating osteoporosis and muscle loss—conditions affecting millions worldwide.

A newer breakthrough—bioprinting in microgravity—has taken biology to new heights. On Earth, printed tissues often struggle to stay intact because of gravity. However, in space, where gravity isn't a factor, scientists can layer delicate "bio-inks" made of living cells and hydrogels to create functioning tissues. Since 2016, the ISS has been home to a BioFabrication Facility where researchers have successfully printed blood vessels, cartilage, and even small patches of heart tissue. By 2023, they've grown simplified human organ models, called organoids, to better understand diseases. These pioneering experiments are bringing us closer to the goal of printing full organs—potentially helping to address the global organ shortage, where only a fraction of patients receive the transplants they need.

From Skylab to the ISS, space experiments have also revealed surprising ways human cells and bacteria act in microgravity. Some bacteria become more virulent in space, while human immune defenses tend to weaken. Interestingly, cancer cells in microgravity naturally form 3D structures that look much more like real tumors compared to many Earth-based models. This speeds up research in cancer, immune therapies, and age-related conditions like Alzheimer's. The ISS also has the Microgravity Expansion Platform, where companies test new medicines and regenerative medicine techniques. They are exploring ways to control inflammation, repair tissues, and heal cells, opening the door to improved treatments for patients Earth.

Immunity, Heart Health, and Radiation

Studying the human body in space opens up incredible opportunities to learn things we simply can't observe on Earth. It gives us unique insights into how our immune system, heart, and ability to withstand radiation work. On the International Space Station, where everything feels almost weightless, each cell acts a little differently. Microgravity influences how our immune system operates, how our hearts adjust, and how cells respond to radiation.

In space, the human immune system tends to weaken, making astronauts more susceptible to infections. Microgravity changes how cells behave in several ways. T cells—our body's first responders—become sluggish, producing fewer cytokines, which are signaling proteins that help coordinate immune responses. Other immune cells, like macrophages and neutrophils, become less effective, and regulatory T cells can sometimes over-suppress the system. All these changes together diminish the body's ability to repair and defend itself.

Interestingly, this immune slowdown mirrors accelerated aging. Cells act as if they have aged faster, which reduces the body's capacity to repair itself and fight off diseases. By understanding this quick "space aging," scientists are better equipped to find ways to strengthen immune health, not only in space but also for aging communities on Earth.

By exploring these immune changes, researchers are working on targeted strategies to boost T-cell function or minimize harmful immune suppression. This includes innovative approaches that use natural immune-activating compounds. The insights gained from this research could lead to better treatments for autoimmune diseases, more effective vaccines, and new therapies for infectious illnesses.

Space also offers a unique window into heart and blood vessel function. On Earth, gravity pulls blood downward, but in microgravity, bodily fluids shift toward the upper body and head. This fluid redistribution changes how the heart pumps and how blood pressure is regulated. For instance, astronauts often experience a "puffy face" and reduced leg volume as blood pools differently. Scientists aboard the ISS use advanced imaging and monitoring to study these changes in detail, seeing how heart muscle adapts and how vessels respond to altered pressure. The findings deepen our understanding of blood pressure regulation and heart function and help inform improved approaches to conditions like hypertension and heart failure.

Unlike on Earth, where our atmosphere shields us from most of the radiation that comes our way, the ISS orbits above much of this protection. Astronauts and experiments in space are exposed to a wide range of cosmic rays and solar particles, which can damage cells and DNA. Scientists study how this radiation affects cellular structures, gene expression, and biological functions, tracking changes with sensitive biosensors and molecular analysis. This research informs not only how to protect astronauts on long-duration missions—such as journeys to Mars—but also improves radiation therapies for cancer patients on Earth. Understanding how cells respond to space radiation helps design drugs that repair DNA damage or boost natural cellular defenses, reducing side effects from radiation treatments. It also guides development of better shielding materials for spacecraft, satellites, and even medical facilities.

By studying how the body reacts when gravity is removed and radiation increases, scientists are rewriting what we know about health and aging. Microgravity research now fuels breakthrough far beyond spaceflight—accelerating drug development, refining medical treatments, and revealing how resilient the human body can be when it leaves its home planet.

Collaborations and Challenges

The pace of space science is accelerating thanks to partnerships among government agencies like NASA, leading pharmaceutical firms such as Merck, Novartis, Amgen, and Bristol Myers Squibb, and innovative biotech companies including Space Tango and SpacePharma. Advanced research modules from companies like Axiom Space and Redwire open new avenues for tissue engineering, organoid growth, and in-space manufacturing. This mix of public and private collaboration is driving groundbreaking discoveries and emerging industries centered on the promise of research in microgravity.

Conducting experiments in space demands large investments, cutting-edge equipment, and rigorous safety measures. Spacecraft size and resources limit the number and complexity of experiments, and samples must survive launch stresses and the harsh space environment. Regulatory agencies such as the FDA are still developing frameworks to approve space-produced drugs and therapies. Transporting biological materials safely between Earth and orbit is another challenge due to temperature changes, radiation, and vibrations. Encouragingly, as commercial space ventures grow and launch costs decrease, these barriers are gradually easing, enabling more frequent and diverse research projects.

Impact on Earth's Healthcare and Technology

Space research is already making a positive impact on medicine and technology here on Earth. Studies conducted in orbit have helped improve cancer treatments and other important medicines by shedding light on how molecules and cells act. Our understanding of bone and muscle loss has led to better care for osteoporosis and muscular conditions. Discoveries about immune and heart health are guiding us toward more accurate diagnostics and preventive healthcare around the world. Technologies that were once only available to astronauts—like advanced drug delivery systems and self-healing materials—are now starting to be used in hospitals and innovative industries. Even in its early stages, space-based tissue and organ printing gives hope to millions waiting for transplants in the future.

As space medicine progresses, the links between orbital experiments and healthcare on Earth will grow ever stronger. Each experiment aboard the ISS helps unravel complex questions about disease, healing, and drug interactions. The knowledge gained will lead to more effective, personalized treatments—and someday could extend human health far beyond our planet.

Better Living Through Chemistry

Chemistry behaves differently in microgravity. On Earth, gravity causes liquids and chemicals to mix and react in ways that often disrupt crystal formation and molecular interactions. But Skylab's research showed how without convection currents—because of microgravity—molecules act differently, allowing purer physical and chemical reactions under controlled conditions.

The Space Shuttle program expanded these studies, investigating processes like combustion, crystallization, and protein folding. Scientists grew protein crystals in space that were clearer and more perfect than Earth-grown ones, improving drug design for diseases. They also created new plastics through polymerization studies—materials now used in medical devices and aerospace.

On the ISS, scientists have pushed these advances further, inventing "smart materials" such as fabrics that regulate temperature in astronaut suits. These materials open new possibilities for industries well beyond space.

Blue Fire

One of the stranger phenomena in space is how fire behaves completely differently without gravity. On Earth, flames rise upward because heat makes air expand and become lighter causing it to float up while the cooler air sinks—like a hot air balloon. This movement of hot air is called *convection*.

But in the microgravity environment of space, flames form perfect, glowing blue spheres that float instead of stretching upward. This behavior was first seen during early space missions in the 1970s, but it wasn't until NASA's groundbreaking FLEX (Flame Extinguishment Experiment) in 2009 and SAFFIRE (Spacecraft Fire Safety Demonstration) in 2016 that scientists really began to understand how fire behaves in space.

Weirdly, it turns out that fires in space burn at lower temperatures and require less oxygen than fires on Earth. Without gravity-driven convection (which also explains the circulation problems in the ISS) to supply oxygen, space flames rely on slow diffusion—oxygen gradually reaches the flame, creating strange, floating blue fireballs. These spheres can last longer than earthly flames, making them both mesmerizing to watch and potentially dangerous inside spacecraft.

Learning how fire works in space has led to important improvements in fire safety on Earth. Engineers have developed new fire suppression systems that work regardless of direction, recognizing that flames don't always travel upward. These advanced systems protect crews aboard submarines, where changing orientation affects how fire spreads.

Aircraft manufacturers have also improved fire detection devices based on space research. Even designers of skyscrapers apply these findings, especially in elevator shafts where fires can behave similarly to those in microgravity. This research shows how studying fire in space helps protect us in challenging environments here on Earth.

Fire's unusual behavior in space also affects how metals mix and solidify. On Earth, gravity causes heavier elements in molten metal to sink while lighter ones rise, creating uneven cooling, structural weaknesses, and impurities in alloys. Convection currents stir the molten mix, leading to swirls and inconsistencies. In space, however, these effects disappear. Molten metals cool evenly and stay perfectly mixed, allowing all elements to bond smoothly and uniformly.

Scientists are exploring the potential of microgravity environments to create innovative alloys that boast incredible heat resistance, flexibility, and strength. These space-made materials are already being tested for demanding applications such as jet engines, which need to withstand very high temperatures, and for surgical implants, which resist corrosion and can last longer inside the body. The possibility of developing these nearly perfect, orbit-forged materials could mean stronger airplane parts, longer-lasting medical devices, more durable electronics, and spacecraft that are both lighter and tougher. It's amazing to think that the strange blue fire and floating metal in space could someday help make our lives safer back home.

Chapter 29

Building in Space

A Flying Industrial Complex

When we hear the phrase "building in space," we might think about astronauts assembling new space stations or working on moon bases with all their high-tech tools, big bulky spacesuits, and giant spanners. But actually, there are many other ways to do things in space!

Manufacturing in space is really revolutionizing the way we create, assemble, and use materials—both beyond Earth and right here at home. From 3D printing, robotic assembly, and orbital factories, things that once seemed like science fiction are now becoming practical, precision-engineered solutions.

Earth's gravity influences almost everything around us. Every building and device has to be sturdy enough to hold itself up and resist forces like shaking or bending. But in space, where microgravity or zero-gravity conditions exist, these limitations vanish—objects float and weight isn't a concern anymore. This fundamentally changes the way materials act and creates exciting new opportunities for designing and making items that simply couldn't be done on Earth.

A big perk of building in space is the freedom from rocket limitations. Here on Earth, every piece has to fit inside a rocket's fairing and endure the violent trip to space. But in orbit, parts can be launched separately or even assembled right where they'll be used. Thanks to 3D printing and robotics, engineers can create huge telescopes, solar panels, and habitats—things that are too big or fragile to launch fully assembled.

Space manufacturing started with human hands, with early missions like Mir and the International Space Station, requiring careful planning and hands-on assembly by astronauts in tough conditions. It was a slow and costly process. But then, automation stepped in. Engineers started exploring new ways to produce materials right in orbit–melting metals, mixing polymers, and testing adhesives to see how materials respond in the unique environment of space.

The leap into 3D printing, often called additive manufacturing, has revolutionized how we create objects. Instead of carving shapes from larger blocks of material, this innovative method builds objects layer by layer from digital blueprints–much like stacking sheets of clay to form a precise and intricate shape. In orbit, this process required some clever innovation. Gravity on Earth helps keep molten plastic or metal steady during printing; in space, those same droplets could float away, creating a challenge. Engineers at Made in Space found a solution by developing printers that control how material is extruded and bonded. Now, astronauts aboard the ISS can print tools, replacement parts, or components whenever needed–turning raw materials directly into useful hardware within just hours. It's like having a tiny, self-sufficient workshop in orbit, saving both costs and space while reducing launch weight.

Robotic assembly quickly became an essential part of space construction. Whether operated remotely from Earth or following automated programs, robotic arms work tirelessly to assemble habitats, satellites, and large structures in orbit. In space, without gravity, robots can build in all directions without the worry of weight, opening up exciting possibilities for innovative designs like expandable space stations or rotating habitats that simulate gravity. A great example of this is NASA's Canadarm2, which is a key robotic system that helps assemble complex space structures.

Materials behave in surprising ways in space! Unlike on Earth, where heavier elements tend to sink, lighter ones rise during mixing, and cooling can be quite uneven, in orbit these processes occur more evenly. That can produce stronger, lighter, and highly reliable alloys and polymers. Experiments have also shown that fiber optics and semiconductors made in microgravity can sometimes exceed Earth-made quality, hinting at faster communications and more durable aerospace parts in the future. Space manufacturing is opening up new possibilities for better materials and technology.

Robotic assembly techniques refined for space missions are already influencing how factories, warehouses, and hospitals operate on Earth, making them safer and more efficient. And the ability to 3Dprint parts on demand, pioneered for spaceflight and rapidly adopted on Earth, is reshaping supply chains and manufacturing around the world.

Earth's orbit is becoming an incredible bustling workshop where robots, printers, and raw elements come together to innovate, craft materials, and construct incredible structures that would be impossible to build on the ground.

As we push the limits of what can be created beyond our planet, these technologies promise not only to drive humanity's exploration of space but also to enhance technology and everyday life here on Earth.

Chapter 30

Space Junk

Space Debris

Space debris–scattered pieces of junk left behind by old satellites, rocket parts, and even tiny fragments of paint–has become a serious, ongoing challenge for the ISS and the astronauts aboard it. The danger of flying space debris was famously shown in the movie *Gravity*, where a speeding cloud of debris smashes into a space station, sending it spinning out of control and putting the crew's lives at risk. While the real ISS is much better prepared, the scene captures just how dangerous these fast-moving objects can be.

Since the very start, keeping the ISS safe from these kinds of threats has been our top priority. Can you imagine, even a tiny speck of paint can hit with the force of a bullet when traveling at orbital speeds–that's just incredible!

To protect against these impacts, spacecrafts are equipped with Whipple Shields–multi-layered armor originally invented by the legendary American astronomer Fred Whipple (1906-2004) in 1947. Over time, NASA's Dr. Burton Cour-Palais (1925-2004) made significant improvements to these shields. When debris hits the outer layer, it shatters and disperses, preventing damage to the inner hull and keeping the crew and vital equipment safe.

However, the problem became more serious in the 1980s, when aerospace engineer Donald Kessler realized that if two pieces of debris collide in orbit, they break apart into many fragments. These fragments can then strike with even more objects and initiate a dangerous chain reaction. This process, now known as the Kessler Syndrome, means that the amount of space debris can increase quickly and make space travel more dangerous for everyone. Because of this risk, Kessler and other engineers focused on strengthening spacecraft shields to accommodate the increased number of high-speed objects in orbit. Today, monitoring space debris and maintaining protective systems are ongoing responsibilities for engineers and astronauts, all to keep the ISS and its crew safe.

As the ISS races around Earth at an astonishing 28,000 kilometers per hour, these advanced shields safeguard both the crew and the station's equipment from potentially catastrophic impacts. The station's safety is further enhanced by advanced tracking software developed under Dr. Nicholas Johnson, who led NASA's Orbital Debris Program Office. His team created systems capable of predicting and avoiding collision paths in the increasingly crowded space environment.

The threat from space debris is very real: microscopic paint flecks can strike with the force of a speeding bullet, and larger objects such as bolts or tiny meteoroids could potentially puncture station modules. To tackle this, engineers developed the Debris Assessment Software (DAS) in the 1990s, under Dr. Johnson's leadership. This software models potential collision risks and has grown more sophisticated in recent years by integrating machine learning to improve predictions.

Remarkably, these space-safety innovations have influenced life on Earth. The same multi-layered shield principles that protect the ISS have inspired the development of improved bulletproof vests and stronger armor for military vehicles. Although space debris tracking technology doesn't directly translate, it has inspired collision warning systems in aviation. Perhaps most impressive is how technology originally designed to monitor the ISS's structural health is now used by engineers to inspect buildings, bridges, and other vital infrastructure on Earth. These systems detect potential problems early, helping to make our world safer for everyone.

Clearing Debris

Scientists and engineers have made significant progress in cleaning up space debris. The European Space Agency's ClearSpace-1 mission aims to test a robotic spacecraft to capture old satellites with mechanical arms that basically work like a space garbage truck.

Other projects are exploring giant nets, magnetic capture systems, and laser pulses that nudge debris into lower orbits, where it will safely burn up on reentry.

Today's satellites are now engineered with end-of-life disposal in mind. Many come with small thrusters that push them into lower orbits when their missions are over helping them reenter and disintegrate in Earth's atmosphere in a controlled way. Special systems also vent leftover fuel or gases to prevent explosions caused by pressure buildup, while improved battery management helps prevent leaks and fires as satellites age.

Some teams are trying out new ways to clean up space debris. They use strong magnets to pick up metallic pieces without touching them directly. Others are using laser beams that push debris into safer orbits, kind of like a stream of water guiding leaves downstream.

Another interesting innovation is the "drag sail"—a big, lightweight sheet that unfurls to catch traces of the upper atmosphere. It helps slow down the satellite's descent, much like a parachute, making decommissioned satellites fall back to Earth and helping reduce the chances of creating long-lasting space junk.

Avoiding Debris

Managing space traffic has evolved into a global network of tracking systems and collision prediction technologies. The US Space Force serves as the primary space traffic controller, operating an extensive network of ground-based radars, optical sensors, and radio telescopes that monitor over 30,000 objects continuously. These tracking stations, spread across the globe, work together like a worldwide surveillance system for space.

Private companies have revolutionized debris tracking capabilities. LeoLabs, for instance, operates a network of phased array radars that can detect objects as small as a golf ball at distances up to 1,000 kilometers. Similar innovations from companies like ExoAnalytic Solutions use optical telescopes to track objects in geosynchronous orbit, about 36,000 kilometers above Earth. These advanced systems process millions of observations daily, creating detailed catalogs of space objects and their trajectories.

The collision warning process has also become more sophisticated. Modern tracking systems use artificial intelligence and machine learning to predict potential collisions up to several weeks in advance. When a potential collision is detected, satellite operators receive detailed conjunction data messages (CDMs) that include probability calculations and recommended avoidance maneuvers. These warnings give operators time to plan and execute careful orbital adjustments while minimizing fuel consumption.

Despite these advances, serious challenges remain. Current systems struggle to track objects in very high orbits or objects smaller than ten centimeters. The increasing use of autonomous collision avoidance systems also raises new questions about liability and decision-making in space traffic management. As more satellites launch and debris accumulates, the need for more accurate tracking and prediction technologies becomes increasingly critical.

Why This Matters

The space debris problem has wide-reaching consequences that many people might not fully realize. It's not just about shielding astronauts and satellites—this issue directly affects our daily lives right here on Earth. The satellites we rely on provide vital services we often take for granted. Our phones use GPS signals from satellites to help us navigate through cities and find our way on road trips. Weather forecasters rely on satellite data to monitor dangerous storms and give us accurate forecasts that keep us safe. When we make international calls, browse the internet, or watch satellite TV, we're depending on clear paths through space for those signals to reach us.

These satellites also serve a number of emergency and scientific research. When natural disasters happen, first responders use satellite data to coordinate rescue missions and evaluate damage. Scientists use satellites to observe climate change, study ocean currents, and gain a better understanding of our planet. Many nations also depend on satellites for national security and monitoring potential threats.

What makes this challenge so tricky is that, even though space seems endless, we're only using certain orbital paths around Earth that are becoming busier with both functioning satellites and dangerous debris.

If we don't act, the consequences could be severe. Future generations might find it impossible to launch new satellites or spacecraft due to the dangerous clutter of debris. Unlike trash on Earth that eventually breaks down, space debris can stay orbiting for hundreds of years before it naturally falls back and burns up in the atmosphere.

The good news is, solutions exist—but they require cooperation from countries worldwide. We need to work together to establish better space traffic management, kind of like how air traffic is coordinated. Investing in innovative technologies to clean up already existing debris is also essential. International rules need to be stronger, with clear consequences for irresponsible behavior. Both private companies and governments should be responsible for their activities in space, including the entire lifecycle of their satellites and equipment.

Protecting Earth's orbits is a shared responsibility, similar to caring for our oceans and atmosphere—it's about teamwork, clear rules, and global stewardship. Space might seem boundless, but our usable orbits are limited, valuable, and getting more crowded every day. Keeping them clean is vital—not only to protect our astronauts but also to secure the satellites that guide us and support future exploration.

Chapter 31

Future Labs

The Post-ISS Future

For more than twenty years, the International Space Station (ISS) has been a shining symbol of teamwork across the globe, serving as a frontier laboratory and a cozy home away from home in space. Since its first module launched in 1998 and it became continuously inhabited in 2000, it has been an incredible platform for humanity to learn how to live and work beyond Earth–hosting thousands of experiments that reshaped our understanding of biology, physics, and technology.

Of course, time is catching up with the ISS. Most of its modules–launched between 1998 and 2011–are starting to show their age, and plans are underway to deorbit it around 2030. While it will be a bittersweet moment to see the ISS come down to Earth, it also opens the door to an exciting new era of space habitats that are smaller, smarter, and more diverse.

Instead of a single, large, government-operated orbital station, the future seems to be heading towards multiple smaller stations run by private companies. Think of it as moving from one huge university campus to a lively city filled with specialized research labs, business hubs, hotels, and creative spaces–all orbiting Earth. This new vision promises more innovation, flexibility, and access for everyone.

Many companies are leading this exciting charge. Axiom Space, working closely with the aerospace manufacturer Thales Alenia Space, is slowly building its vision of the future, one step at a time. They started by attaching new, stylish modules to the ISS, designed by the renowned architect Philippe Starck to have elegant interiors reminiscent of luxury hotels. These modules will initially extend the ISS but are planned to detach later, becoming their own independent Axiom Space Station. This gradual, modular approach combines proven technology with new levels of comfort and style, making long-term space living more appealing and inviting.

Meanwhile, Blue Origin has teamed up with Sierra Space, Boeing, and Redwire to create Orbital Reef—an exciting new floating business district in orbit. Think of it as the "Silicon Valley" of space, where companies can rent offices, laboratories, and manufacturing spaces. Orbital Reef aims to welcome a diverse range of visitors: scientists conducting experiments, filmmakers capturing stunning movies in zero gravity, entrepreneurs eager to innovate, and wealthy tourists looking for unforgettable space vacations. This station is designed to be versatile and user-friendly—a bustling hub for commercial activity.

Starlab, a collaboration between Nanoracks, Voyager Space, and Airbus, takes a focused scientific approach. It will be a single-module station dedicated to research and space manufacturing. Notably, it will feature inflatable rooms that expand once in orbit, providing more space without adding extra weight for launch, and advanced robotic arms to assist with experiments and maintenance, ensuring smooth operation with less human effort. This innovative design will make Starlab a powerful platform that pushes the boundaries of science and technology.

Vast Space, the newest player in the field, dreams big with plans to build the first artificial gravity space station called Haven-1. Unlike traditional stations where occupants float in microgravity, Haven-1 will spin slowly to generate centrifugal force, mimicking Earth's gravity. This twist could greatly improve comfort and help reduce health issues like muscle loss or bone weakening during long-term missions. Vast Space is partnering with SpaceX to launch Haven-1 aboard the Starship rocket, opening a new chapter in human space habitats.

These upcoming stations are different from the ISS in many ways. They are smaller, more specialized, and smarter. Many will use artificial intelligence to automate daily tasks, making life easier for crew members and boosting efficiency. Advanced 3D printing will enable on-site manufacturing of tools and parts, reducing the need for costly resupply missions. Some stations might even include sections with artificial gravity, easing physical strain on their occupants.

And it's not just astronauts who will benefit from these stations. Scientists will rent precious lab space to conduct cutting-edge experiments that can benefit life on Earth and beyond. Pharmaceutical companies may develop new medicines, while manufacturing firms can create materials that can't survive Earth's launch forces. Students and educators will send experiments into space, inspiring the next generation of explorers. Artists, chefs, filmmakers, and tourists will visit these stations too—pushing the boundaries of creativity and experiencing life beyond our planet.

Having many stations brings a wonderful diversity of focus and expertise. One station might focus on creating new materials for space and Earth industries, while another might specialize in growing food in space for future missions to the Moon or Mars. There's also a station that could serve as a training ground and waypoint for lunar explorers. This kind of specialization is like having multiple expert labs instead of just one large, general-purpose building, which really encourages innovation and progress across many fields. These exciting new outposts aren't just the successors to the ISS— they represent the emergence of a vibrant new space economy. By the 2040s, Earth's orbit could be bustling with labs, hotels, and robotic factories, each with its own unique mission and character. The next time you look up at the night sky and see a station gliding overhead, you'll be witnessing one of our greatest experiments: life itself among the stars.

Chapter 32

Giant Spinny Things

Simulating Gravity

Wouldn't it be cool to float freely in space for a while where every movement is effortless? This sensation of weightlessness is fun for brief moments, but it comes with significant challenges for human health and the practicalities of living and working in space for extended periods.

To address these challenges, scientists and engineers have developed ways to create artificial gravity–simulated weight–to protect astronauts and enable activities that require Earth-like conditions.

Why Artificial Gravity Matters

Gravity shapes life. On Earth, it guides the flow of blood, strengthens muscles and bones by giving them something to push against, and influences countless biological and physical processes. Without it, everything changes. Astronauts' muscles weaken, bones lose density, and fluids drift toward the head, straining the heart and blurring vision. Even at the cellular level, growth patterns and organ functions shift in unexpected ways. Over time, the human body simply begins to forget what "down" feels like.

That's why creating artificial gravity–simulated weight–is a crucial step toward keeping humans healthy on long missions in orbit and beyond. Artificial gravity helps maintain bone and muscle strength, keeps fluids circulating naturally, and even reduces the psychological stress of constant weightlessness, where eating, sleeping, and moving can feel awkward and disorienting. It gives astronauts a sense of "normal," something their bodies and minds deeply need.

More than comfort, it could transform how we work in space. Many industrial processes on Earth, like mixing liquids, forming metals, or building complex structures, rely on gravity's steady influence. In microgravity, these processes behave unpredictably. With artificial gravity, factories in orbit could produce materials that are stronger, purer, or entirely new—things impossible to make on Earth.

Artificial gravity is more than a convenience; it's the foundation for long-term life and industry in space. It keeps humans strong, machines efficient, and the dream of living among the stars within reach.

Creating Gravity Through Rotation

Creating artificial gravity through rotation is one of the most practical methods. Imagine a gigantic spinning bicycle wheel, with astronauts living and working along the rim—it's a vivid idea that helps us understand the concept. As it spins, their bodies tend to move straight ahead, while the ring's structure provides a centripetal force, keeping them moving in a circle. This makes them feel pressed toward the outer edge, just like feeling gravity on Earth. The rotation needs to be steady and slow enough to feel natural, and engineers are already exploring designs that could offer more comfortable, Earth-like gravity.

In the future, spacecraft and space stations might have sections that rotate around a central axis. As they spin, people inside would feel pushed outward, creating a sensation similar to gravity. The faster the spin or the larger the circle, the stronger this artificial gravity becomes. But spinning too quickly could cause dizziness or nausea because of the Coriolis effect, which makes your inner ear feel like you're twisting when you move your head or limbs. To keep things comfortable, engineers are likely to keep rotation rates between two and four revolutions per minute and use rotating sections that are tens of meters wide for a smooth, gravity-like experience.

Building such a rotating habitat in space is a major engineering challenge. Large rotating rings or cylinders require a lot of materials and complex structures. Some designs include giant space wheels, like O'Neill cylinders, which are enormous carousels kilometers wide, spinning slowly to simulate Earth-like gravity on their inner surfaces.

Others prefer smaller options, such as centrifuge modules attached to non-rotating spacecraft, where astronauts can spend limited time adjusting to gravity before moving into microgravity zones. These modular setups offer flexibility for experiments needing zero gravity and habitats needing artificial gravity.

In the future, spacecraft could use linear acceleration—gently pushing forward—to create a sensation of gravity, similar to the feeling you get when a car accelerates quickly and presses you into your seat. Accelerating at around 9.8 meters per second squared, which is similar to Earth's gravity, would give astronauts that familiar feeling inside the spaceship. However, keeping this up for long journeys would need very efficient propulsion systems, which are still being developed.

Artificial Gravity as a Gateway to the Stars

Space agencies and companies envision a future where artificial gravity becomes a routine feature of space habitats. Large rotating space stations could serve as homes and research centers, spinning slowly to create comfortable environments for extended stays. NASA, for example, has studied designs with rotating rings connected to central hubs that provide varying gravity levels. This allows astronauts to experience Earth-like gravity in living quarters while scientific equipment operates in microgravity zones.

For deep space voyages, particularly Mars missions, spacecraft may include rotating sections where astronauts live and work during months-long journeys. This design would minimize health risks from microgravity exposure and help maintain physical fitness.

Artificial gravity is key to making long-term space habitation safe, comfortable, and productive. By mimicking Earth's gravity, we can protect astronauts from physical decline caused by microgravity, enable complex manufacturing of new materials, and create habitats that feel more like home.

As technology advances reduce costs and improve engineering, spinning stations and rotating spaceships will transition from concept to reality, transforming human space exploration. Soon, artificial gravity will become standard wherever humanity ventures—from orbiting space hotels and factories to colonies on the Moon, Mars, and beyond—making life among the stars feel as natural as life on Earth.

Chapter 33

Getting Home

The Challenge of Returning from Space

Returning to Earth can be one of the most challenging parts of space travel. When a spacecraft comes back, it needs to slow down from incredible speeds of 28,000 kilometers per hour while enduring the violent physics of atmospheric reentry. As it plunges into the Earth's atmosphere, the thin air thickens fast, compressing the craft and creating shock waves that are hot enough to melt metal—over 1,600°C (3,000°F).

To protect itself, the spacecraft uses a thermal protection system, which is a special layered heat shield. This shield works by absorbing, redirecting, and shedding heat through methods like ablation (where part of the shield material gradually burns away), radiation (releasing heat into space), and conduction (spreading heat across the shield to prevent hot spots). Successfully managing these thermal processes is crucial to keeping the spacecraft and its crew safe from being incinerated during reentry.

Bow Shock

In the 1950s, H. Julian Allen (1910-1977) and his team made a discovery that forever changed how we explore space: it turns out that blunt-shaped spacecraft are much safer for reentry than the sleek, pointed ones people previously believed were the best. While streamlined shapes work great for airplanes flying in the atmosphere, the incredible speeds involved in reentry—over tens of thousands of miles per hour—a blunt body creates a bow shock wave in front of the spacecraft, acting like a shield.

When a spacecraft smashes into the atmosphere at supersonic speeds, the air in front of it compresses with incredible force. It's similar to quickly smacking your hand into water–water splashes around your fingers, right? The same thing happens with air: molecules collide, forming a thick wall of compressed gas called a shock wave. With its blunt shape, this shock wave stays away from the spacecraft's surface, forming a protective zone. This detached bow shock wave acts like an invisible thermal shield, much like holding a shield a few feet away from a roaring fire. Even though the shock wave heats the air to extremely high temperatures, most of this superheated plasma stays separated from the spacecraft by a cooler layer of gas, thinner than a credit card. This cushion helps push away the harshest heat, keeping the vehicle–along with the astronauts inside–protected. Whether this process leads to a safe landing or a disaster depends on how well we understand and manage these forces.

Thanks to Allen's discovery, engineers moved from sleek, pointed capsules to blunt-bodied spacecraft and started using ablative heat shields. These shields are made from special materials designed to sacrifice themselves on purpose: as they heat up, their surface chars, may melt, and then break apart into hot gas and tiny fragments that are swept away, carrying heat off in a process called ablation. By allowing the outer layers to burn, erode, and peel away, ablative shields protect most of the intense heat from reaching the metal structure and the astronauts inside.

The Space Shuttle, despite having a more complex shape, still featured a blunt underside that helped create a bow shock wave for added heat protection. Its thermal protection system included heat-resistant tiles and carbon-carbon composites that shielded the orbiter from the super-hot air caused by shock heating. Modern spacecraft follow this helpful principle; for example, the SpaceX Dragon capsule uses an ablative shield that gradually burns and erodes away during reentry, taking heat with it as the material is stripped off. The newest spacecraft, like SpaceX's Starship, combine traditional metal heat shields with active cooling systems, highlighting how our understanding of thermal management continues to grow and improve.

Reentry Blackout

The journey back to Earth is a carefully choreographed crash landing. The spacecraft first needs to orient itself just right, with its heat shield facing outward to guard against the intense heat. The angle of reentry is absolutely critical. If the angle is too steep, the vehicle will be exposed to crushing g-forces and extreme heating, while too shallow an approach means the spacecraft risks skipping off the atmosphere like a stone skimming across a pond.

As the spacecraft plunges back into Earth's atmosphere from orbit, it travels through the thick layers of air, facing forces that can be up to five times Earth's gravity. The surrounding air becomes compressed and heated to thousands of degrees, stripping electrons away from atoms and creating a sheath of ionized gas called plasma. While visually spectacular, this glowing cocoon can also block radio waves, interrupting communication with mission control in a brief and tense interval called communications blackout or reentry blackout.

Usually, this blackout starts shortly after the spacecraft enters the denser parts of the atmosphere and ends once it slows down enough for the plasma to fade away. The blackout can last anywhere from a few moments to several minutes, depending on factors like the size, shape, speed, and entry angle of the spacecraft. During NASA's Mercury, Gemini, and Apollo missions, blackouts typically lasted between three to six minutes. For example, Apollo 13 endured an unusually long blackout of about six minutes because of its atypical reentry trajectory.

Modern engineering has made this reentry process far safer and more predictable. Onboard computers constantly refine the reentry path, and networks of sensors track temperatures, structural loads, and vehicle condition in real time.

Communications have improved as well. Systems such as NASA's Tracking and Data Relay Satellite System (TDRSS) can route signals through satellites positioned outside the plasma sheath, shortening blackout periods for certain missions. Even so, the physics remains unchanged: any vehicle returning at orbital or interplanetary speeds must push through that envelope of superheated air.

As we'll see next, materials used in heat shields have benefited from decades of research in materials science and thermal dynamics, leading to innovations that go beyond spaceflight–these same technologies now help firefighters and improve aircraft design.

The successful return of spacecraft to Earth is one of humanity's greatest achievements in applied physics. It shows how our understanding of thermodynamics, fluid mechanics, and aerodynamics comes together to solve an incredibly complex challenge. Each reentry is a testament to human ingenuity, showing how we can harness the fundamental laws of nature to bring our astronauts safely home.

Chapter 34

Space Age Materials

Space's Unforgiving Hostility

Space challenges every material we send beyond Earth. Extreme temperatures, radiation, high-speed impacts, and strange chemical reactions mean ordinary materials are not enough. These harsh conditions push engineers to invent lightweight, strong materials that can survive where almost nothing else can.

In the vacuum of space, there is no atmosphere to soften temperature swings. Surfaces can bake in sunlight and then plunge into deep cold within minutes. These brutal thermal cycles would crack or warp typical materials, so spacecraft rely on special composites and coatings that can flex, expand, and contract without breaking down.

Beyond Earth's protective magnetic field, spacecraft face constant attacks from radiation, ultraviolet light, solar flares, and cosmic rays, all of which slowly damage materials at the molecular level. Engineers have responded by creating tough, resilient substances that stay reliable for years. At the same time, and as we heard before, tiny bits of orbital debris race around Earth at about 28,000 kilometers per hour, hitting with bullet-like energy, so designers build smart layered shields that can absorb and spread out those impacts.

Chemical erosion adds another layer of difficulty. In low Earth orbit, ultraviolet light creates highly reactive atomic oxygen that eats away surfaces atom by atom. The vacuum also causes materials to "outgas," releasing vapors that can contaminate delicate instruments. To combat these problems, specialists develop coatings that resist erosion, block atomic oxygen, and remain stable without any air around them.

Together, these challenges have sparked a broad family of space-age materials: shape-memory alloys that unfold structures in orbit, ultra-light aerogels that provide excellent insulation, carbon-fiber composites stronger than steel but much lighter, and ceramic tiles that withstand the intense heat of reentry. The search for better space materials continues as missions grow longer and venture farther, and these very cool solutions for spacecraft are increasingly finding their way into everyday life back on Earth.

Tough Stuff

In their quest to explore space, engineers have been searching for materials that are both incredibly strong and lightweight. Steel was simply too heavy, and aluminum wasn't quite durable enough. This challenge sparked the creation of revolutionary new materials that have transformed not just space travel but also many aspects of our everyday lives.

Titanium and Alloys

Titanium first appeared in materials science back in 1791 when William Gregor discovered unusual black crystals in Cornwall. Although early on scientists recognized its potential, extracting pure titanium was impossible for the next 150 years because it reacts so strongly with oxygen.

The big breakthrough came during World War II, when William Kroll developed a clever two-step process: first, converting titanium ore into titanium tetrachloride with chlorine gas; then, using magnesium to carve out pure titanium metal. This Kroll process finally made titanium production feasible.

NASA soon saw the amazing qualities of titanium: steel-like strength at just half the weight, excellent resistance to corrosion, and stability even under extreme temperatures. It became a vital material for Apollo lunar modules, Space Shuttle parts, the International Space Station (ISS), Mars rovers, and the James Webb Space Telescope.

On Earth, titanium also made a huge impact in medicine thanks to its biocompatibility, making it perfect for implants like artificial joints and dental work. Its resistance to corrosion is ideal for ocean exploration equipment and its incredible strength-to-weight ratio led to the development of advanced sporting goods and consumer products.

Building on the strength of pure titanium, engineers have created specialized alloys such as Ti-6Al-4V, which are even more capable for tough applications. These alloys are roughly forty percent lighter than steel but just as strong, capable of withstanding extreme conditions—from Venus's crushing ninety atmospheres of pressure and surface temperatures reaching 470°C, to the vacuum of space.

These versatile alloys are found in countless applications, including Apollo lunar modules, Mars rovers, Voyager probes, and modern spacecraft like SpaceX's Crew Dragon. On Earth, they help improve jet engines, medical implants, chemical processing equipment, and high-performance sporting gear.

Carbon Fiber

Carbon fiber changed the game in materials science during the 1950s. Made from super-thin strands of pure carbon, it starts as a synthetic polymer—usually polyacrylonitrile—which is spun into fibers and then heat-treated at temperatures soaring up to 3,000°C. This process aligns carbon atoms into incredibly strong crystalline structures.

These fibers are woven into sheets and combined with resins to create composites that marry carbon's powerful strength with the flexibility of resin. Companies like Union Carbide and scientists like Dr. Roger Bacon pioneered carbon fiber technology to produce materials that are stronger than steel but at a fraction of the weight.

Beyond aerospace, carbon fiber has helped create lighter, more fuel-efficient vehicles, improved sporting equipment, and provided durable yet lightweight options for everyday products—from laptop cases to eyeglass frames.

Thermal Chaos

Managing temperature in space can be quite challenging, but it's essential for keeping spacecraft and their crews safe. Without Earth's atmosphere, spacecraft face dramatic temperature swings that can damage equipment and pose risks to crew safety. To cope, they need to shield themselves from both freezing cold and scorching heat. For example, when in orbit, one side might be warmed to about +120°C, while the shaded side cools down to around −150°C, which can cause materials to warp, crack, or fail.

Reentry conditions are even more demanding. At orbital speeds, the air in front of the spacecraft turns into superheated plasma, capable of damaging or destroying many materials. The nose of the Space Shuttle, for instance, reached temperatures high enough to melt most metals.

Space Blankets

The shiny gold "space blankets" used by marathon runners and hikers actually have their roots in spacecraft technology. They're based on Multi-Layer Insulation (MLI), which was developed to protect spacecraft from extreme temperature changes. Created back in the early 1960s at NASA's Marshall Space Flight Center, MLI stacks ultra-thin layers of aluminized Mylar with mesh spacers in between. Piling up twenty to thirty layers can reflect up to about ninety-seven percent of radiant heat, while the vacuum between layers helps reduce heat transfer and keeps spacecraft temperatures more stable.

During the Apollo missions, NASA wrapped the Lunar Module in MLI to handle temperature swings of roughly +260°F in sunlight and −280°F in shadow. Today, spacecraft like Voyager and the International Space Station still rely on MLI for thermal control.

In 1964, Dr. Robert Montgomery adapted this technology for everyday use by creating inexpensive five-layer "space blankets." Instead of trapping air like a thick coat, they work more like a thermos, reflecting your own body heat back toward you.

Modern versions of these blankets incorporate advanced materials such as aerogels—mostly empty space by volume—and ultra-thin metal coatings. Today, variants of MLI are used in many places, like keeping blood and vaccines cold in hospitals, maintaining the temperature of food deliveries, reducing building energy costs by forty percent, and providing lightweight warmth for campers and outdoor athletes.

Teflon

Teflon's journey begins in 1938 when DuPont scientist Roy Plunkett opened a gas cylinder and discovered a surprising white waxy coating inside. This discovery led to the development of a new kind of plastic—Polytetrafluoroethylene (PTFE)—which is known for being incredibly slippery, highly resistant to chemicals, and capable of withstanding a wide range of temperatures from about −240°C to +260°C. Basically, it's a long chain polymer made of carbon atoms surrounded by fluorine. Since carbon and fluorine form some of the strongest bonds in organic chemistry, PTFE is very stable and resistant to chemical reactions.

During World War II, this durable and inert material proved essential in the Manhattan Project, lining valves, gaskets, and pipes that transported highly corrosive uranium compounds without leaking. Over the years, its remarkable properties made it perfect for space exploration too. NASA used Teflon as electrical insulation for cables, low-friction seals, and protective layers that could withstand launch vibrations, vacuum conditions, and extreme temperatures.

In the Apollo era, Teflon was woven into "Beta cloth," the white outer fabric of spacesuits and some spacecraft surfaces. It helped resist fire, micrometeoroid pitting, and lunar dust. Variations of Teflon-coated fabrics have also flown on the Space Shuttle and now line parts of the International Space Station, where PTFE continues to appear in seals, wiring, and thermal protection.

Back on Earth, the influence of space-grade Teflon inspired new products. In 1954, French engineer Marc Grégoire used PTFE to coat his wife's pans, launching the very first non-stick cookware and turning Teflon into a household staple. Today, this same space-tested material is found in medical implants, sutures, Gore-Tex® waterproof membranes, architectural roof fabrics, and many industrial coatings and seals. testaments to a polymer that once helped astronauts reach the Moon.

Lightweight Ceramic Insulation Tiles

In the mid-1970s, Lockheed engineers developed a new material for the Space Shuttle—lightweight ceramic tiles that could withstand 2,300°F during reentry while remaining light enough for spaceflight.

These amazing tiles, developed in 1976, used silica fibers arranged to create a structure that was ninety-four percent empty space. They were so light they could float on water, yet they resisted extreme reentry heating; you could hold one in your bare hand while the opposite face glowed at around 2,000°F. The Space Shuttle carried more than 24,000 of these tiles as its thermal shield.

Today, innovative ceramic insulation is making an impact worldwide by enhancing industrial furnaces, catalytic converters, firefighting equipment, and building insulation. It also protects aircraft "black boxes." Space agencies continue to push the boundaries with advanced heat shields designed for the most extreme environments. For example, the Venera probes to Venus faced scorching temperatures and high pressures, while the Huygens probe to Titan used ablative shields that slowly burned away to manage heat, supported by composites that maintained their strength at high temperatures. These exciting technologies also contribute to the development of reusable spacecraft, like SpaceX's Dragon capsule, helping to make space travel more sustainable and efficient.

Reinforced Carbon-Carbon

Reinforced Carbon-Carbon (RCC) is quite remarkable in the world of materials science! It's specially engineered to preserve much of its strength and stiffness even at extremely high temperatures. Developed through collaboration between NASA and industry partners like LTV Aerospace in the early 1970s, RCC can handle scorching temperatures around 1,700°C while still maintaining its shape and integrity.

The creation of RCC is a multiyear process. It begins with layers of carbon-fiber cloth soaked in phenolic resin, which are then heated in large pressure vessels to over 2,000°C. This heat turns the resin into carbon, firmly bonding with the fibers. Once formed, the carbon–carbon structure is coated with silicon carbide and special sealants to protect it from oxidation at high temperature.

The Space Shuttle heavily relied on RCC for its most critical hot spots—the nose cap and wing leading edges—where temperatures could reach nearly 1,600°C during reentry. Following the Columbia accident, engineers chose not to replace RCC outright. Instead, they improved inspection methods, on-orbit checks, and protective coatings to better detect and manage any damage.

Today, similar carbon-carbon composites are found in high-performance areas like Formula One brake discs, experimental hypersonic aircraft components, and parts of industrial furnaces designed to withstand intense heat. Aerospace companies are continually working to refine RCC-like materials, making them lighter, stronger, and more affordable for space missions and high-temperature uses on Earth.

Aerogel

Aerogel began with a bar bet in 1931. Chemistry professor Samuel Kistler wagered that he could remove the liquid from a jelly without letting it collapse. After months of experiments, he pulled it off using supercritical drying—heating and pressurizing the gel until the liquid turned to gas without destroying the structure.

The result looked like solid smoke—a nearly transparent material that's 99.8 percent empty space. This structure gives aerogel extraordinary properties: it can support 4,000 times its weight, provide excellent insulation against extreme temperatures, and can absorb significant energy from fast-moving projectiles. A small brick-sized piece weighs less than a feather, earning it the nickname "frozen smoke" and recognition amongst the world's least dense solids ever made.

Safety Comes First

Space exploration demands extraordinary safety measures. In the endless expanse of space, where temperatures can get extremely hot or cold and radiation is very strong, even small mistakes can lead to big problems. Over time, engineers have developed innovative solutions to tackle these challenges, creating technologies that protect astronauts and even improve our daily lives here on Earth.

To shield astronauts' eyes from the bright glare of the sun, NASA engineers in the early 1960s made visors coated with a thin layer of gold. This shiny coating reflected harmful rays while still letting astronauts see clearly. Today, this same idea helps firefighters, workers who weld, and even office workers—appearing in face shields, car headlights, sunglasses, and window coatings that keep heat and energy under control.

Protecting the rest of the body required different kinds of innovation. After the tragic Apollo 1 fire in 1967, NASA completely overhauled how it handled fire safety in space, adopting nonmelting, charforming fabrics such as Beta cloth for spacesuits and spacecraft interiors. Later research on how flames behave in microgravity—forming floating spheres instead of rising—deepened scientists' understanding of fire and helped refine space-station fire-safety designs. Similar principles now show up in many flame-resistant materials used in firefighter suits, building insulation, and even children's pajamas.

Radiation was another invisible danger. Outside Earth's magnetic field, astronauts face constant bombardment from cosmic rays. Engineers started using dense, heat-resistant metals like tantalum and new semiconductors such as gallium nitride (GaN), which can handle radiation while staying lightweight. Today, GaN powers many things—from electric cars and 5G systems to important medical equipment.

Collisions were just as risky. Traveling at about 17,500 miles per hour, even a tiny speck of paint can puncture a spacecraft. As mentioned earlier, engineers solved this with Whipple shields—several thin layers placed apart to slow down, break, and trap incoming particles. The same idea now keeps people safe inside military armor, car crumple zones, and sports helmets.

Even the cushions astronauts sit on led to new innovations. In 1966, NASA created memory foam, a special type of polyurethane that spreads out pressure and absorbs shocks to protect pilots and astronauts during launch and landing. It quickly became common in airplane seats, mattresses, prosthetics, and countless comfort products on Earth.

From gold visors to Whipple shields, every safety improvement inspired by space travel has made life safer below the atmosphere. The relentless effort to protect astronauts has given us tools that protect firefighters, patients, drivers, and everyday people—showing that the quest for safety in the most dangerous environment has also helped make Earth a safer place to live.

Weird Substances

Space exploration has given birth to some weird stuff that seems almost like it's straight out of a science fiction story. These futuristic materials can change shape, repair themselves, and even bend light—all thanks to humanity's drive to explore and thrive beyond Earth.

A wonderful example is shape-memory alloys, which are metals that "remember" their original shape. Discovered accidentally in 1959 by metallurgist William J. Buehler, a nickel-titanium sample had exhibited an extraordinary trait—it could return to its pre-set shape when heated. NASA later used these alloys to deploy solar panels and antennas that unfold themselves once warmed in orbit. Today, you'll find the same technology in medical stents, eyeglass frames that bounce back after bending, and even earthquake-resistant buildings that flex and reset after tremors.

Just as bizarre are smart fluids—liquids filled with tiny iron particles that can stiffen instantly when exposed to magnetic fields. Originally developed to stabilize spacecraft and dampen vibrations, these magnetorheological fluids now smooth the suspensions of luxury cars, help prosthetic limbs adapt to walking speed, and even safeguard skyscrapers and bridges such as China's Dongting Lake Bridge by hardening in milliseconds during seismic shifts.

Carbon nanotubes, discovered in 1991 by Japanese scientist Sumio Iijima, are incredibly tiny tubes made of carbon atoms that are rolled up like chicken wire. These fascinating structures have opened up exciting opportunities in various fields, showcasing the incredible potential of nanotechnology. They're so small that 50,000 could fit across a single hair, yet strong enough that a pencil-thick bundle could theoretically lift a truck! These "tiny tubes" conduct electricity better than copper and transfer heat more efficiently than diamonds, revolutionizing technologies from flexible electronics and water filters to medical sensors and ultra-light sports gear. Perhaps even more exciting is their potential to someday make space elevators a reality.

Inspired by the resilience of human skin, self-healing materials contain microscopic capsules filled with liquid "healing agents" that burst open when damaged, sealing cracks and regaining much of their strength. Since Professor Scott White invented them in 2001, they've become part of our everyday lives—think car paints that repair scratches, phone screens that mend themselves, and bacteria-infused concrete that seals itself when exposed to water.

There is a lot of excitement around metamaterials—ingenious "meta metals" engineered with tiny patterns smaller than the wavelengths of light or sound they aim to control. These amazing structures can bend electromagnetic waves in extraordinary ways, sometimes even making objects seem invisible at certain frequencies. Developed first in 2001, metamaterials now play a role in satellite antennas, smartphone cameras, and state-of-the-art medical imaging devices. NASA is exploring them for ultra-light radiation shields, while scientists around the world dream up everything from better earthquake protection to more efficient solar panels.

Materials that can reshape themselves or fluids that harden on command give us a glimpse into the future of innovation. As our space adventures challenge our creativity, we often find that the most unusual inventions from above become some of the most valuable here on Earth.

Chapter 35

Pristine Workshops

How to Build Space Stuff

Creating materials for space began in the early days of NASA and the Soviet space program in the 1950s when engineers had to build things that could survive in an environment with no air, extreme temperatures, and constant radiation. This led to the development of specialized manufacturing facilities that were cleaner, more precise, and more controlled than anything seen before.

These facilities evolved from simple clean workspaces in the Mercury program to today's advanced manufacturing centers used by companies like SpaceX, Blue Origin, and Boeing. Each advancement in space technology brought new requirements for how things needed to be built.

Pristine Rooms

Have you ever seen a team assembling a spacecraft, all dressed in white gowns, gloves, and masks, and wondered why such precautions are necessary?

NASA, other space agencies, and aerospace companies create these clean room environments to build and assemble the most sensitive parts of spacecraft like satellites, orbiters, landers, and rovers. Even the tiniest speck of dust or contamination can cause serious malfunctions or failures in space systems. Making sure the environment is extremely clean helps protect delicate electronic circuits, sensors, optical equipment, and scientific instruments, so they work perfectly during years of exposure to space's harsh conditions.

In a clean room, every effort is made to keep out dust, microbes, and tiny airborne particles that, though invisible, could harm the complex technology on spacecraft. For example, NASA's Mars rovers are assembled in clean rooms to ensure they work well on the rugged Martian surface. The same goes for cameras, telescopes, and other precise instruments, which need spotless conditions to produce clear images and data because even a tiny particle can degrade their performance.

Clean rooms are also essential for missions that gather samples from space, like the NASA Genesis spacecraft, which collected particles from the solar wind. These sterile environments keep Earth-based contaminants—such as dust, bacteria, or spores—away from the extraterrestrial materials, keeping the scientific value intact. They are equally important for making spacesuits and life-support systems, where sterility and reliability are crucial for astronaut safety during missions.

Besides assembling spacecraft in clean rooms, every part is carefully tested, cleaned, and packaged to stay free from contaminants during transport to launch pads and beyond. This thorough care helps meet planetary protection rules, which are designed to prevent Earth microbes from accidentally traveling to other planets and interfering with the search for alien life.

The technology behind clean rooms actually started long before space exploration. It all began with Willis Whitfield, a physicist at Sandia National Laboratories, who in 1962 solved a major problem in making tiny mechanical parts. Before his work, attempts to create dust-free workspaces often failed because workers stirred up more dust than filtration systems could handle. Whitfield's solution was to push filtered, clean air from the ceiling down through tiny holes in the floor, creating a constant "invisible waterfall" of clean air flowing through the space. This airflow carries dust and contaminants away, keeping the environment thousands of times cleaner than before—even cleaner than typical hospital operating rooms.

At first, many experts were skeptical about Whitfield's clean room concept; during one meeting, a Japanese manufacturer famously called it "magic." However, companies like RCA saw the immediate benefits and quickly adopted the technology. The timing was perfect, especially as the computer industry grew, since microchip manufacturing required dust-free spaces to avoid costly defects. Major companies like IBM and Texas Instruments soon built clean rooms based on Whitfield's design, which became the standard for decades and was officially codified in Federal Standard 209.

As more industries recognized its importance, Whitfield's clean room technology expanded beyond electronics. Pharmaceutical companies use clean rooms to produce medicines without contamination. Hospitals rely on them for sterile surgical environments. NASA used this technology to assemble highly sensitive spacecraft parts that could fail if exposed to particles. Today, clean room technology is the backbone of countless industries—from Intel's advanced chip factories, which are thousands of times cleaner than Whitfield's original design, to manufacturers of pacemakers, artificial joints, contact lenses, and even food production facilities that depend on clean environments to keep products safe and fresh.

Vacuum Testing

Since the early days of the space race, recreating the extreme conditions of space here on Earth has been a vital part of helping us prepare spacecraft for their journeys. One of the most important milestones in this journey came in 1962, when NASA engineer Norman Appold led the design of Chamber A at the Johnson Space Center in Houston. Completed in 1965, this enormous vacuum chamber was built to test spacecraft and equipment under the challenging conditions of space long before they ever launched.

Chamber A is a marvel of engineering. It's as tall as a fourteen-story building, with aluminum-lined walls surrounding a space large enough to fit a four-story building inside. Its original goal was to test hardware for the Apollo missions, making sure they could withstand the intense vacuum and extreme temperatures of space. In 2012, the chamber received a major upgrade to prepare for testing the James Webb Space Telescope including advanced systems to push temperatures even lower—down to an incredible -250°C, only a few dozen degrees above absolute zero—the coldest temperature possible and a limit we can never quite reach (remember the Kelvin scale and the Third Law of Thermodynamics?).

The chamber creates this environment with powerful vacuum pumps–technology pioneered by Wolfgang Gaede–that remove 99.99999% of the air inside. To achieve the extreme cold, it is lined with special shrouds cooled by liquid nitrogen and liquid helium. Around the clock, highly precise sensors keep a close watch on every detail of the chamber's environment to ensure perfect testing conditions.

Though this technology was originally developed to prepare spacecraft for space, it has found many valuable applications here on Earth. Electronics companies like Samsung and LG use vacuum chamber designs inspired by space testing facilities to stress-test their TV and phone displays, making sure screens stay reliable despite temperature changes and pressure differences during shipping. Medical device manufacturers like Steris and Getinge use vacuum chamber technology to sterilize sensitive surgical instruments and equipment. The food industry also benefits– companies like FoodSaver have adapted vacuum sealing methods, originally inspired by space chamber technology, to keep food fresh for longer.

Even automotive and construction industries have embraced aerospace vacuum testing innovations. These technologies help create better weather-testing chambers that simulate extreme temperature and pressure conditions for vehicles and building materials, improving their reliability and durability in real-world conditions.

Pressure Cookers

An autoclave is a powerful machine, much like a giant pressure cooker, used to apply intense heat and pressure to materials inside a sealed chamber. This process helps special composite materials– layers of fibers and resin–to bond and harden into strong, lightweight structures. In the space industry, autoclaves play a crucial role in making spacecraft parts that need to withstand extreme conditions like launch stresses, vacuum, temperature variations, and radiation.

The autoclave works by sealing materials inside a chamber where temperatures can go above 400°C (750°F) and pressure can often reach ten times normal atmospheric pressure. This highly controlled process cures composite materials, turning them into tough, reliable components while keeping weight to a minimum—key for efficient space travel where every kilogram counts. Big aerospace companies like Boeing use enormous autoclaves that can cure entire airplane wings in a single go, like for the 787 Dreamliner, while Airbus uses similar facilities for large sections of their aircraft.

The history of autoclave technology in aerospace goes back to the late 1940s, when engineers at Wright-Patterson Air Force Base started adapting pressure vessels to create stronger, more dependable structural materials. The first autoclave specifically designed for aerospace manufacturing was built in 1953 by United Aircraft Corporation (now part of United Technologies). Around the same time, the foundations of modern composites were being laid. Dr. Roger Bacon's early work on carbon whiskers showed how strong carbon could be, and in the 1960s, British scientists William Watt and Leslie Phillips expanded on this. They developed a method to heat synthetic fibers such as polyacrylonitrile (PAN) to produce high-strength carbon fibers—five times lighter than steel yet remarkably strong. This breakthrough made it possible to create the lightweight, high-performance composites that autoclaves now cure so effectively.

Today, autoclaves are essential in aerospace and many other industries. Besides space and airplane parts, they are used to produce high-performance carbon fiber composites found in tennis rackets, wind turbine blades, racing bicycles, golf clubs, and even laptop cases. Medical prosthetics also benefit from autoclave-cured composites that are strong, lightweight, and compatible with the human body. The construction industry increasingly relies on carbon fiber composites to reinforce concrete bridges and buildings, helping them withstand earthquakes.

From their beginnings in aerospace manufacturing to their widespread use today, autoclaves continue to be a vital part of advanced production—enabling stronger, lighter, and more durable materials that push the boundaries of exploration and bring innovation to everyday life.

As advanced as these materials are, their real value only emerges through the precision of the manufacturing and machining that shape them.

Flawless Machinists

Precision manufacturing in space really owes a lot to the innovative work done at the Massachusetts Institute of Technology (MIT) in the late 1950s. There, John T. Parsons and Frank L. Stulen pioneered the invention of Computer Numerical Control (CNC) machining. They faced the challenge of creating aircraft parts more consistently and accurately than what human machinists could manage, so they developed a method using early computers to control cutting tools. This breakthrough allowed machines to follow precise mathematical instructions, shaping metal with incredible accuracy far beyond manual methods.

When NASA kicked off its space program, it quickly became obvious that traditional manufacturing wasn't precise or reliable enough for spacecraft parts. Every gram of weight and every tiny imperfection could jeopardize a mission. Thanks to MIT's Servomechanisms Laboratory, supported by the Air Force, the first practical Numerical Control (NC) system was created. It used punched tape–resembling a player piano roll but for guiding machine instructions–to steer cutting tools.

The real game-changer came in the 1960s as CNC machines became fully digital. Companies like Cincinnati Milacron and Fanuc started making machines that could produce parts with extraordinary precision–accurate to just 0.001 inches, about the width of a human hair. Such precision was crucial in space, where even the smallest flaw could lead to failure.

Today's CNC machines are even more advanced and adaptable. Manufacturers like DMG MORI in Japan and Haas Automation in California build machines that can work with nearly any material–from soft aluminum to tough titanium. They can move cutting tools along five or six axes, guided by sophisticated computer programs that simulate and perfect every cut before the machine begins working. This technology makes it possible to create complex shapes that would be impossible to carve by hand.

CNC technology has had a huge impact beyond space exploration. Hospitals depend on CNC-machined parts for artificial hips and dental implants. Jewelers use tiny CNC machines to craft intricate designs from precious metals. Car companies shape perfectly-fitting engine parts with giant CNC machines. Even artists use CNC to create sculptures that would be too difficult or time-consuming to carve manually. Originally developed to meet the tough standards of space, CNC machining has become a fundamental part of modern manufacturing, showing how innovations from space exploration continue to improve life here on Earth.

Printing On-Demand

3D printing in space begins with three visionary entrepreneurs: Aaron Kemmer, Jason Dunn, and Mike Chen. In 2010, they founded Made in Space, which is now part of Redwire Space, with a bold dream: to create the first manufacturing facility in orbit. These Silicon Valley engineers saw that space missions were limited by having to pack everything they might need for their journey. What if, they wondered, astronauts could simply print what they needed, when they needed it?

After years of testing in NASA's "Vomit Comet" aircraft that simulates zero gravity, the team developed a special 3D printer that could work without gravity, pulling the melted plastic down. Unlike regular 3D printers that rely on gravity to help layer materials, their printer used special mechanisms to control the flow of plastic in weightless conditions. Back in 2014, they achieved what many thought impossible—their compact printer, roughly the size of a microwave, managed to print its very first part right on the International Space Station. This wasn't just a cool science experiment. The printer showed that astronauts could fix equipment by printing new parts instead of waiting months for the next supply rocket. If something broke on the space station, they could potentially download the design from Earth and print a replacement part within hours. The printer could work with several types of strong plastics, including the same material used to make Lego bricks.

The technology they developed has found surprising uses back on Earth, especially in medicine where similar high precision 3D-printing techniques are now used by medical implant companies. The careful control needed for printing in zero gravity turned out to be perfect for printing delicate replacement joints and dental implants. Surgeons can now scan a patient's body and print implants that fit perfectly, leading to better outcomes and faster healing. What started as a solution for astronauts has become a breakthrough for patients on Earth, showing how space technology often finds unexpected ways to improve life on our planet.

SECTION 4: EXPLORING SPACE

Reaching for the Stars

Exploring space is one of humanity's most exciting adventures. It's incredible to think that our journey has taken us from those cautious first steps on the Moon to dreaming about settling other worlds. This collection of knowledge feels like a guided tour of our solar system, showing how we're learning to thrive in some of the harshest places imaginable and creating pathways through the vastness of space.

As we move from the Moon to Mars and beyond, we're tackling some pretty tough challenges—developing life support systems that work in a vacuum, designing habitats that protect us from radiation, finding ways to mine on alien landscapes, and uncovering water in the most unexpected places. Every solution not only helps us reach further into space but also brings innovations back to Earth that improve our daily lives.

Space exploration reflects our human ingenuity and curiosity about our place in the universe. It's about learning how the insights we gain among the stars can help us take better care of our own planet. As we explore the solar system, we're reaching outward while also viewing Earth with fresh eyes and a deeper appreciation.

In this section, we celebrate our cosmic ambitions—from setting up temporary camps on alien worlds to building permanent homes, creating interplanetary transportation systems, and pushing the limits of human exploration. It's all about expanding what we believe is possible and maybe, one day, calling other worlds home.

Chapter 36

Our Alien Neighbor

Landing on the Moon

Landing on the Moon presents challenges unlike anything experienced on Earth. Since the lunar surface has no atmosphere to cushion a descent—no winds, no air resistance, no parachutes—the conditions are harsh and unforgiving. One side of the Moon bakes under intense sunlight, while the other remains in deep cold. Even though these conditions are predictable, there's no margin for error, as the lunar environment offers no help to slow down or adjust a spacecraft's fall.

Every part of the descent needs to be planned to ensure safety and success. The spacecraft's engines will play a crucial role in providing just the right amount of braking to counteract lunar gravity, which is about one sixth of the Earth's. It's important that these engines modulate their power smoothly, follow a detailed, carefully mapped trajectory, and make the most of their limited fuel supply. Achieving this delicate balance depends on perfectly syncing the thrust, timing, and navigation to work together seamlessly.

The early attempts at lunar landing showed just how tough this challenge really was. Between 1959 and 1976, the Soviet Luna program achieved an impressive series of milestones: the first impact on the Moon, the first photographs of the far side, the first soft landings, the first samples returned, and the first rover exploring another world. The United States soon joined in with the Surveyor missions (1966-1968), which made America's first soft landings. Surveyor proved that spacecraft could land safely and that the lunar surface was able to support their weight. It also studied the soil, tested landing techniques, and laid important groundwork for future human missions.

The idea that eventually enabled Apollo was actually proposed decades earlier. Back in the 1920s, Ukrainian engineer Yuri Kondratyuk suggested a method where a smaller lander would detach from the main spacecraft in lunar orbit for the landing. He kept his notes hidden during the turmoil of war and never lived to see their significance. In the early 1960s, NASA engineer John Houbolt revived this concept–called Lunar Orbit Rendezvous (LOR)–and spent years convincing skeptics that it was the best way to meet President Kennedy's ambitious deadline. In this plan, the Command and Service Module would stay in orbit while a separate Lunar Module descended to land.

Apollo 11 proved the concept to the world. While Michael Collins orbited above in the Command Module Columbia, Neil Armstrong and Buzz Aldrin guided the Lunar Module Eagle toward the surface. They faced computer alarms, low fuel, and the need to steer away from a dangerous boulder field–all with seconds of fuel left–before safely landing. That moment marked humanity's very first controlled landing on another world.

A lunar landing involves three key stages: setting up orbit, starting the descent, and using engine thrust to achieve a soft touchdown. Without an atmosphere, the engines must stay in almost constant control to manage the fall.

Since the Apollo missions, many countries have made great strides in lunar landing technology. China's Chang'e program introduced hazard-avoidance systems. In 2019, Chang'e-4 became the first to land on the far side of the Moon, and in 2020, Chang'e-5 returned samples. India's Chandrayaan-3 reached the Moon's south pole in 2023. Japan's 2024 SLIM mission demonstrated precise landing using advanced image recognition to match the surface below to onboard maps.

Now, a new chapter is unfolding. NASA's Artemis program aims to build a lasting human presence on the Moon, working with commercial partners and international agencies. SpaceX's Starship will deliver cargo and crew with its powerful Raptor engines, and Blue Origin's Blue Moon lander will help develop infrastructure. The

Lunar Gateway, orbiting in a near rectilinear halo orbit, will serve as a staging point for surface missions and future journeys to Mars. New, heavy-duty rovers will expand how far crews can travel and what they can build.

Together, these missions mark a shift from brief visits to the Moon toward establishing a permanent presence—extracting resources, conducting science, developing technologies for Mars, and inspiring a new generation of explorers. Humanity is beginning to expand outward, reaching for new horizons.

Surviving the Lunar Extremes

The Moon is our closest neighbor in space, yet its environment is very different from anything we experience on Earth. Without an atmosphere or magnetic field to protect life, it faces extreme temperature swings—from scorching highs of about 120°C during the lunar day to freezing lows near -130°C at night. These tough conditions influence every decision behind human exploration and settling there.

To deal with these challenges, engineers are creating advanced life-support systems that aim for nearly complete recycling of air and water. These systems use a mix of filters, chemical reactors, and biological processes to recover resources rather than discard them. Such approaches could one day support lunar habitats and also provide practical benefits for people in remote communities or disaster-hit areas on Earth, where clean water and safe living environments are just as important.

Temperature control calls for equally creative ideas. Scientists are exploring phase-change materials that soak up and release heat as they melt and solidify, multilayer insulation to cut down heat loss, and reflective coatings that keep temperatures steady. Many of these methods are already used in high-performance buildings and industrial systems on Earth, helping to boost efficiency in tough climates.

Radiation protection is one of the Moon's biggest hurdles. Without a magnetic field or atmosphere, solar and cosmic particles hit the surface directly. Researchers are trying various protective strategies: burying habitats under lunar soil, building walls from regolith-based bricks, and using hydrogen-rich materials that better block energetic particles than metal. These strategies could protect future lunar crews and also improve shielding for satellites, medical facilities, and workplaces with high radiation on Earth.

Even power–the backbone of all systems–needs smart solutions. A promising idea is to place solar panels near the lunar poles, where high ridges called "peaks of eternal light" get almost nonstop sunlight, helping to solve the problem of night-time darkness. Scientists are also looking into nuclear power and advanced energy storage to keep habitats and equipment running through the two-week-long lunar night. These energy storage solutions address similar challenges faced in Earth's polar regions and remote places lacking stable power grids.

Together, these technologies reveal how the Moon's harsh environment is inspiring new ideas in life support, thermal management, radiation shielding, and energy systems. The same solutions that help humans survive on the lunar surface also strengthen our resilience here on Earth–turning space exploration into a powerful driver for improving life at home.

The Gravity Challenge

The Moon's gravity is about one-sixth that of Earth's, creating a physical environment that's quite different from what we experience on our planet. Objects weigh much less, but their mass stays the same, and this small change influences just about everything in lunar life–from how our bodies respond to how machinery behaves.

For people, reduced gravity means rapid adaptation. Without Earth's constant pull downward, muscles don't have to work as hard and can weaken over time. Meanwhile, bones that usually get stronger through bearing weight can lose density–around one to two percent each month on the International Space Station. Fluids shift upward, changing how the heart and circulation work, reducing the heart's workload, and gradually leading to deconditioning.

To help address these challenges, engineers and scientists have developed specialized solutions. Advanced resistance exercise machines use vacuum cylinders, flywheels, or electromagnetic systems to mimic Earth-like forces since traditional weights don't have much effect in low gravity. The Gravity Loading Countermeasure Skinsuit adds tension from shoulders to feet to simulate Earth's pull and help maintain bone and muscle health. Researchers are also exploring medicines that could slow down bone loss or support muscle growth, and future artificial gravity systems using centrifuges could offer short "gravity therapy" sessions for astronauts on long missions.

Lunar gravity also impacts machinery in important ways. Many systems designed on Earth depend on gravity to keep fluids moving, joints engaged, or parts seated properly. In lunar conditions, liquids flow more slowly and surface-tension effects become more pronounced, complicating fuel management, cooling, and hydraulics. Moving parts relying on weight to maintain contact might slip or lose friction, and stability is a concern too: machines might weigh less, but their mass and momentum stay the same, making vehicles more likely to tip over when they accelerate or turn.

To tackle these issues, engineers design lunar equipment with wider bases and lower centers of gravity. Mechanical systems are reengineered so they can work without relying on Earth's gravity. Fluid handling uses pressurization and capillary action instead of simple gravity-fed methods. On Earth, testing often involves harness rigs and partial-gravity simulators to mimic one-sixth g and see how equipment behaves when weight changes but inertia doesn't.

Interestingly, solving these lunar challenges can lead to unexpected benefits back on Earth. Exercise devices originally made for Moon habitats are now inspiring new rehab tools for people recovering strength or mobility. Techniques for maintaining bone density are aiding osteoporosis treatments, and cardiovascular studies from spaceflight are helping develop therapies for heart patients. Exploring how bodies and machines adapt to different gravity environments helps us deepen our understanding of physiology and engineering, which benefits both space missions and healthcare here at home.

Dusty Razor Blades

Lunar regolith–also known as Moon dust–poses some of the most persistent challenges in lunar exploration. Unlike Earth soil, which has been smoothed over billions of years by wind and water, lunar dust is made up of sharp, jagged particles created by continual meteorite impacts. Under a microscope, it looks like countless tiny shards of glass. Solar radiation gives these particles an electric charge, making the dust cling tightly–almost magnetically–and allowing it to stick to spacesuits, instruments, and mechanical joints, even sneaking into the smallest openings.

Apollo astronauts faced these issues firsthand. Dust scratched their visors, wore down suit parts, clogged zippers and seals, and even caused breathing trouble when it entered the Lunar Module. Commander Gene Cernan famously called lunar dust "one of our greatest inhibitors" to smooth operations.

Today, engineers are working on stronger solutions to tackle what might be the Moon's most underestimated environmental hazard. NASA and university partners are developing the Electrodynamic Dust Shield–a system that uses oscillating electric fields to push dust away from surfaces like solar panels, camera lenses, and spacesuit visors. Researchers are also looking into lotus-inspired hydrophobic and oleophobic coatings that make surfaces smoother and less likely to attract dust. NASA's xEMU suit development team is including tougher fabrics, redesigned bearings and seals, and suitports designed to keep dust out of habitats.

Mechanical systems also need their own protective measures. Hardened seals, dust-resistant lubricants, and fully enclosed joints are being created to reduce wear on robotic arms, wheels, drills, and hinges. ESA and DLR's LUNA facility, among others, test equipment in conditions that mimic real lunar soil, which is abrasive and electrostatically charged.

Interestingly, studying lunar dust has already brought benefits back here on Earth. Semiconductor manufacturers now use dust-repelling surface treatments inspired by space research. Medical device makers apply similar coatings to reduce contamination risks. Mining and industrial workers in desert areas are adopting dust-tolerant mechanisms first tested on lunar robots, improving safety and reliability.

While the Moon's razor-sharp, electrically charged dust can be an intimidating challenge, the innovations developed from studying it are helping create new technologies. These advances not only improve life and industry on Earth but also prepare us for future adventures on the lunar surface.

Artemis

NASA's Artemis program signifies an exciting new chapter in our journey to explore the Moon. Unlike the Apollo days focused on quick landings, Artemis emphasizes creating a lasting presence, with the help of international partners and innovative collaborations.

At the core of this effort is the Lunar Gateway, a small yet strategically situated space station that will orbit the Moon in a special path called a Near-Rectilinear Halo Orbit (NRHO). This elongated, lopsided orbit loops high over one lunar pole and sweeps low over the other, using the gravity of Earth and the Moon to stay steady while conserving fuel. It stays in constant touch with Earth and provides long passes over the poles, making it an ideal base for landers headed to the south pole—a region packed with scientific opportunities and resources.

Artemis isn't just about a single spacecraft or agency; it's a global teamwork effort, featuring companies like Blue Origin with its Blue Moon lander to deliver supplies, Dynetics with versatile cargo platforms, and Northrop Grumman bringing propulsion expertise from its Cygnus spacecraft. Innovators like Astrobotic are developing autonomous landing tech that will help spacecraft touch down safely with minimal human intervention.

All these collaborations capture the spirit of Artemis: a worldwide, cooperative effort driven by shared goals and technological progress. The aim isn't just to visit the Moon again, but to build a permanent presence—creating habitats, harvesting lunar resources, advancing scientific research, testing new tech for future Mars missions, and paving the way for explorers of the next generation as humanity explores further into the Solar System.

Artemis opens the door for us, offering the transport routes, landers, partnerships, and systems that make reaching the Moon possible in a sustainable way. But getting to the surface is just the beginning. The real challenge starts when astronauts step outside the lander and encounter the lunar environment head-on.

To stay, we need places that shield, support, and protect us. This brings us to the next frontier: how to build a home on a world that wasn't originally meant for us.

Lunar Habitation

Building a home on the Moon is an exciting challenge that pushes engineers to explore environments far more extreme than anything on Earth. Without air, magnetic protection, or shield from intense radiation, and with temperatures swinging hundreds of degrees during a lunar day, designing habitats that can withstand these conditions calls for innovative solutions in materials, life support, construction, and power. Many of these innovations also benefit us here on Earth.

Modern habitats use multiple protective layers to ensure safety. Outer shells are made from cutting-edge materials that block radiation and solar particles, while interior systems control pressure, temperature, humidity, and oxygen levels. Together, these systems keep the living environment stable, even when the outside surface heats up like an oven or plunges into deep cold.

A major line of research is turning the Moon itself into a construction site. Engineers are experimenting with transforming lunar regolith into bricks, tiles, or even fully 3D-printed structures using methods like microwave sintering or sulfur-based binders. These materials are especially valuable because they reduce the need to transport building supplies from Earth and naturally shield inhabitants from radiation and micrometeorites. While regolith-based shells can outperform many traditional materials in blocking dangerous particles, they must be designed to withstand the large temperature swings and thermal stresses.

Power is another crucial piece of the puzzle. Since the Moon takes about two weeks to rotate from daylight to darkness, solar power alone isn't enough. Solar panels thrive during the long lunar day, but surviving the two-week night requires advanced energy storage. Researchers are exploring hybrid systems—at the poles, the "peaks of eternal light" receive almost constant sunlight, making them ideal for solar stations. Small modular nuclear reactors are also under study as consistent power sources that don't depend on sunlight, and innovative storage solutions like regenerative fuel cells could help balance energy supply during day-night transitions.

With these combined strategies—durable habitat structures, in-situ construction techniques, and versatile power systems—scientists are laying the groundwork for sustainable lunar living. As they develop these technologies, they're also creating new tools that benefit us here on Earth: stronger building materials, better radiation shielding, and reliable off-grid power systems for remote communities. The Moon is becoming a testing ground not just for surviving off Earth but for finding smarter, more sustainable ways to live everywhere.

Lunar Mining

The development of systems that can tap into the Moon's own resources is one of the key concepts in lunar exploration. These in-situ resource utilization (ISRU) technologies are a blend of mining tools and chemical plants: they extract water from lunar soil and rocks, then turn it into vital materials for life support and mission operations. Essentially, such equipment could find frozen deposits, melt and purify them through electrolysis, and create drinking water—or split the molecules to produce oxygen and rocket fuel, opening up all sorts of possibilities for future lunar missions.

Scientists have confirmed the presence of water ice on the Moon, though it is not spread across the surface. Instead, most of it lies within Permanently Shadowed Regions (PSRs), particularly at the poles. These craters never see sunlight and remain extraordinarily cold—below about -150°C—allowing ice to remain stable for millions of years. Estimates suggest there may be billions of kilograms stored within these frigid pockets. While there may be traces of water elsewhere, the PSRs hold the concentrated deposits that future missions will rely on. Importantly, no spacecraft or astronaut has yet physically accessed or sampled this ice.

Lunar Mining Challenges

Extracting resources on the Moon is quite a challenge, very different from mining on Earth. Lunar dust is a tricky problem: its sharp, glass-like particles, energized by the Sun, cling to equipment, sneak into mechanisms, and wear down moving parts. The Moon's lower gravity makes designing machinery even harder. Vehicles behave in unexpected ways: they don't sink as much into the soil, maintaining traction is tougher, and equipment can tip easily because less weight doesn't mean less inertia. So, heavy machinery needs to be redesigned from the ground up.

Power reliability is another big hurdle. With lunar nights lasting about two weeks, standard solar power just can't keep mining operations going all the time. We need alternative energy sources—like nuclear power, advanced batteries, or hybrid systems—to keep things running through the long lunar nights.

There's also limited hands-on experience with drilling in lunar conditions. The Apollo missions only took shallow core samples, the deepest being around three meters during Apollo 17. Industrial-scale drilling has never been tested on the Moon, so we're uncertain how equipment will perform in vacuum, extreme cold, and abrasive lunar regolith.

Operating remotely adds another layer of difficulty. Without humans nearby to fix problems, systems need to be incredibly reliable or capable of diagnosing issues and repairing themselves. The communication delay of about 1.3 seconds each way may seem small, but it's enough to make real-time control challenging. Dust, temperature swings, and radiation can shorten equipment lifespan, so we need highly durable designs.

Finally, the logistics of lunar mining are tough. The Moon has no roads, processing plants, or transport infrastructure. Building all this takes a lot of investment upfront. The Moon's reduced gravity affects how materials flow during transport and processing, and the abrasive dust speeds up wear and tear on vehicles and conveyor belts.

Changesite & Helium-3

In 2022, China's Chang'e-5 mission discovered a new lunar mineral called Changesite (Y)—a tiny phosphate crystal that's only about one-tenth the width of a human hair. It is the sixth new mineral ever found on the Moon, and China is the first to identify one. While scientists are still exploring its practical uses, this mineral provides valuable insights into the Moon's geological history.

Another resource that's capturing a lot of attention is helium-3—a rare isotope on Earth but found in trace amounts across lunar soil. Helium-3 could be a promising fuel for advanced nuclear fusion, potentially generating large amounts of clean energy with minimal radioactive waste. Estimates suggest there might be around 1.1 million metric tons embedded in the lunar regolith. However, the concentration is extremely low—roughly twenty to thirty parts per billion—so producing just one kilogram would require processing about 150 million kilograms of regolith, which is about sixty Olympic-sized swimming pools of material.

There are still major challenges to overcome. Fusion reactors that could use helium-3 are not yet developed, since current efforts focus on more achievable fusion methods. Also, lunar mining at an industrial scale would require heavy, durable machinery, reliable power sources, and technologies capable of operating in the Moon's extreme cold, radiation, and vacuum—capabilities that are still being developed.

Despite these hurdles, interest in lunar resource extraction is growing rapidly. This work is pushing forward innovations in robotics, materials processing, autonomous systems, and extreme-environment engineering. In fact, these advancements are already helping on Earth, powering remote communities and operating industrial facilities in tough conditions.

This new race isn't just about water, minerals, or helium-3. It's about building the expertise needed for sustained activity beyond our planet. No matter what economic value these resources may eventually hold, the technologies developed during this journey could be just as transformative—shaping a new era of human exploration both on the Moon and right here at home.

Testing Ground for the Future

So we need air, water, power, mining, protection from extreme temperatures and radiation—and of course, a way to grow food!

The challenges sound overwhelming, but here's good news: every solution we invent to survive on the Moon also makes life better on Earth. The same technologies designed to recycle air and water could help cities manage waste more efficiently. Radiation shields built for astronauts might protect patients in cancer treatment centers. And experiments in lunar farming could inspire more sustainable ways to grow food in harsh environments on our planet.

The Moon may be 384,000 kilometers away, but in many ways, it's our closest testing ground for helping us find new ways to sustain life out there, but right here on Earth.

Chapter 37

Becoming Martians

Getting To Mars

Timing is really key when we send spacecraft beyond our Earth. For trips to Mars or Venus, engineers must wait for specific planetary alignments. When it comes to Mars, the best position happens when Mars is opposite the Sun in our sky, and the two planets are quite close. This moment is called opposition. But the ideal time to launch is usually just before or after this, which is what we call the Mars launch window.

This window only opens every twenty-six months, and the best times are often just a few days or even hours long. Getting the timing right in space is more than just launching at the right moment. It's about targeting a moving destination across millions of kilometers and arriving with enough fuel to steer, make adjustments, and land.

Most Mars missions follow a Hohmann transfer orbit, which takes about seven to eight months. This path traces an ellipse around the Sun, touching Earth's orbit on one side and Mars' on the other, timed so the spacecraft meets Mars when it reaches that point. Although slower than some faster options, this method uses less fuel—important when carrying heavy payloads. Nearly all successful missions to Mars, from the early Mariners to Perseverance, have used this route.

But new, faster options are being developed. Innovations like nuclear-thermal engines and high-power ion drives could cut the trip down to three or four months. Shorter trips mean less radiation and less wear and tear on the spacecraft, which is great. However, they also bring new challenges, like higher speeds that cause more heat during entry into the thin Martian atmosphere. Future missions will need stronger heat shields and even more precise navigation as these faster systems become available.

Payloads

Sending payloads to Mars involves three major challenges: escaping Earth's gravity, placing the spacecraft on the correct interplanetary path, and using propulsion efficient enough to carry useful cargo all the way to Mars.

Reaching orbit requires a speed of roughly 28,000 kilometers per hour. Only after achieving orbit can a spacecraft perform the additional burns needed to climb away from Earth's gravity well. Powerful engines and large fuel reserves support this, and companies such as SpaceX have built vehicles precisely for these demands.

The trans Mars injection maneuver follows next. This carefully timed burn places the spacecraft on its trajectory toward Mars. Conventional chemical engines can only burn efficiently for short periods, so mission planners are studying more efficient propulsion for future heavy cargo missions, including nuclear thermal and ion systems that deliver far higher propellant efficiency over long durations.

Landing

Landing is the most demanding phase of a Mars mission. As a spacecraft approaches the planet at about 21,000 kilometers per hour, it meets an atmosphere that is thin yet fierce. There is not enough air to slow the vehicle, but there is enough to compress violently in front of it, creating a glowing layer of plasma hotter than 1,500°C. The heat shield protects the structure beneath by shaping this compressed flow and channeling the heat away.

Deeper in the atmosphere, the air becomes dense enough to deploy the parachutes. These enormous supersonic canopies must open at extreme speeds and in an atmosphere with only about one percent of Earth's pressure. They must be large enough to produce meaningful drag, stable enough to survive chaotic airflow, and strong enough to decelerate heavy payloads.

Parachutes can only slow the descent part of the way. Rocket engines ignite while the vehicle is still travelling faster than the speed of sound. This supersonic retropropulsion pushes exhaust into the oncoming airflow, creating a complex region of shock waves and turbulent plumes. The engines must maintain stability through this chaos and guide the spacecraft towards its landing site.

NASA's skycrane system, used to land Curiosity and Perseverance, manages the final stage. The descent stage slows to a hover, lowers the rover to the surface on cables, and then flies away to crash at a safe distance. SpaceX intends to land Starship differently, using a fully propulsive vertical touchdown similar to Falcon 9 operations on Earth but adapted for Mars' thin atmosphere.

Distance adds another challenge. Radio signals between Earth and Mars can take between four and twenty-two minutes to travel one way, depending on the planets' positions. Entry, descent, and landing must therefore be executed entirely autonomously. This tense period—often called the "seven minutes of terror"—pushes every sensor and onboard computer to its limits.

A successful Mars landing depends on balancing payload mass, propulsion, navigation accuracy, and reliable descent systems—all within narrow launch windows that appear only every two years. Each mission brings us closer to treating Mars not as a distant dream but as a destination we can reach with increasing confidence.

Life on Mars

Mars captures the imagination because it feels familiar: sunsets, seasons, and landscapes sculpted by wind. Yet the familiarity is deceptive. Gravity is only about thirty-eight percent of Earth's, a level that gradually weakens bones and alters how blood moves through the body. The atmosphere is almost pure carbon dioxide and far too thin to breathe. Fine, electrically charged dust infiltrates equipment and reacts with moisture inside habitats, and temperatures swing from mild afternoons to nights well below freezing.

Living on Mars means inhabiting sealed bases that function like spacecraft anchored to the surface. They must generate breathable air, recycle water, and maintain stable temperatures. Water becomes a precious resource, recovered from every possible source—drinking supplies, humidity in the air, even condensation on the walls. Nothing can be wasted.

Because payload mass is limited, Martian systems must serve several purposes at once. Greenhouses are a prime example: plants produce food, absorb carbon dioxide, release oxygen, and provide a vital psychological connection to Earth. Hydroponics keeps them growing efficiently in nutrient-rich water, helping close the habitat's environmental loop while giving crews a daily task that supports emotional well-being.

Sustainable life on Mars demands meticulous management of every resource. Habitats must operate as finely balanced, closed ecosystems where technology and biology work together to support human life in an environment that resembles Earth only from afar.

Learning from Robots

Robotic explorers like Curiosity and Perseverance have paved the way for future human missions. Equipped with cameras, sensors, and a variety of scientific instruments, they traverse rugged terrains, drill into ancient rocks, and brave dust storms—all while sending back valuable data that guides and inspires every step of future exploration.

Dust remains one of the biggest challenges on Mars. Its tiny, chemically reactive particles stick to solar panels, sneak into joints, and interfere with moving parts, gradually lowering the power they generate. These lessons drive the development of dust-resistant mechanisms and consider alternative energy sources for longer missions.

The rovers also teach us how extreme Mars' temperature swings can be—sometimes over 100°C difference between day and night. Electronics and materials need to withstand these harsh conditions while staying reliable for years. Their journeys across dunes, fractured rocks, and steep slopes help identify safe routes and ideal landing spots for future human explorers.

When Curiosity's drill faced a malfunction, engineers had to quickly come up with new ways to operate it remotely, using only the tools already available on board. Such moments highlight the importance of adaptable designs and strong diagnostic skills, especially since human crews won't always have immediate access to spare parts or repair assistance.

Ingenuity, the tiny helicopter carried by Perseverance, made history by achieving the first powered and controlled flight on another planet. That's really impressive, especially considering the thin atmosphere, which is only about one percent as dense as Earth's. Its aerial scouting expanded the rover's ability to explore and has shown that flying vehicles can play a big role in future missions.

All together, these robotic pioneers are helping bridge the gap between distant science and human settlement. They test new equipment, improve operational strategies, and showcase innovations that benefit robotics and remote operations technology not just on Mars, but here on Earth too.

Martian Psychology

The biggest challenge might not be the engineering itself but understanding the human mind. Crews will spend months or even years living in confined spaces with just a few companions, millions of kilometers away from home. Help will be months away, and communication with Earth will have delays long enough to make real conversations difficult.

Being isolated can take a toll on people. Feelings of disconnection, anxiety, or boredom can chip away at mental well-being. Everyday experiences—like walking outside, feeling the wind on your skin, or hearing natural sounds—will be missing, and missing these simple pleasures can be harder than you might think.

To get ready, space agencies study environments on Earth that mimic space living, like Antarctic stations, submarines, and remote desert habitats. These studies show which routines, team interactions, and support systems help people stay healthy during long missions. Some crews do well with shared rituals; others find it tough when privacy or personal time is limited.

Staying in touch with family, even with delays, helps keep emotional bonds strong. Having private spaces gives a much-needed break from constant proximity. Daily routines that balance work, exercise, and leisure can ease psychological stress. Working together on tasks and celebrating small milestones help build team spirit and give each day a steady rhythm.

Getting to Mars will depend just as much on how we support mental health as on the physical engineering. Our habitats and mission plans need to be designed with care to protect mental well-being just like we safeguard against hazards like air issues, water quality, and radiation.

A Practical Approach

The first human bases on Mars will be sturdy, compact shelters designed primarily for survival. These habitats may be built using innovative 3D-printing techniques that fuse Martian soil and rock into strong structures capable of blocking harmful radiation and enduring the planet's extreme temperature fluctuations. Lessons learned from these designs can help us create safer, more energy-efficient buildings here on Earth, inspiring sustainable innovations everywhere.

Living spaces will be efficient and versatile, adapting to various needs. Medical systems will incorporate smart diagnostics and telemedicine technologies, with sensors that send vital data back to doctors on Earth. These technologies already support patients in remote areas, and on Mars, they'll become essential for maintaining health and well-being. Since every kilogram of payload is costly and spare parts are limited, multifunctionality isn't just handy; it's crucial, helping crews keep everything running smoothly with limited resources.

Water is a precious resource on Mars. Habitats will recycle almost all water, building on proven systems used aboard the International Space Station. These innovative water management methods are already making a difference in drought-affected regions and remote communities around the world.

Mars operations will rely on a fleet of orbiters, landers, cargo ships, and support systems. Many of these will serve multiple roles—carrying supplies, functioning as depots, or acting as landers and local power sources—to save weight and boost resilience. Technologies like inflatable heat shields for heavier payloads and AI-powered landing navigation are making rapid progress, opening new possibilities.

Worldwide, agencies and companies are exploring ways to extract resources and produce oxygen from lunar and Martian soils, all while discussing the legal frameworks needed for responsible resource use. Building a future on Mars is about more than just arriving; it's about creating a sustainable, thriving community.

Real missions will make the most of local materials—transforming ice into fuel, soil into shelter, and waste into valuable supplies. Living sustainably on Mars requires resilience, ingenuity, and a deep respect for the limits of an alien world.

Growing Food on Mars and the Moon

Growing food on Mars and the Moon comes with its own set of unique challenges. Despite the weaker gravity—about thirty-eight percent of Earth's on Mars and only around sixteen percent on the Moon—it's still strong enough to guide plant roots downward, unlike microgravity conditions. This makes both planets more suitable for farming than orbiting stations.

The soils here are tricky; Martian soil has perchlorate salts that need to be removed or neutralized, and lunar regolith is sharp and completely lacks organic material. To overcome this, crews will use specially engineered growth media that deliver nutrients precisely.

Radiation levels are significantly higher than on Earth, so habitats and greenhouses need substantial shielding—often made from regolith, water walls, or hydrogen-rich materials.

Crops will grow inside highly controlled environments where temperature, humidity, and light are kept just right. Water will be sourced from extracted ice, purified, and recycled through closed-loop systems. Supplemental LED lighting will also help compensate during Martian dust storms and the Moon's long nights.

Space farming not only supports life-support systems and reduces our reliance on Earth but also brings psychological comfort by introducing green life into these barren landscapes.

Looking Ahead

Exploring life on other worlds shows us the importance of living within our limits. Every discovery in space farming and habitat design opens up new opportunities to conserve resources, adjust to tough climates, and build strong, resilient communities. These valuable lessons naturally come back home, guiding us in areas like food security, caring for our environment, and living sustainably.

The gardens and habitats we develop on Mars will demonstrate that life can flourish even far from Earth and perhaps help all of us to nurture life right here at home.

Chapter 38

Space Power

New Frontiers in Energy

In space, without power, nothing works—not the air we breathe, not the computers that guide us, and not the systems that keep us alive. Unlike on Earth, where electricity is usually just a switch away, space has no power grid or emergency backup. Every satellite orbiting Earth, every robot exploring Mars, and every signal we receive from deep space depends on a reliable power source. In space, there are no wall sockets or repair crews nearby, so each watt must be generated and managed very carefully. Spacecraft need to carry their own power supplies or produce power on the go.

As plans for lunar bases and human missions to Mars get closer to reality, power becomes even more vital. It's not just about turning on the lights but enabling people to live and work hundreds of thousands of kilometers from the nearest power source.

This chapter explores how spaceflight is tackling one of its biggest challenges: keeping everything running smoothly far away from home.

Smashing Atoms

Back in 1938, scientists Otto Hahn and Fritz Strassmann discovered nuclear fission when they bombarded neutrons at uranium, causing it to split into lighter elements. Meanwhile, Lise Meitner, who was working in exile in Sweden, helped explain the science behind it all, showing how splitting atoms could unleash enormous amounts of energy—remember $E=mc^2$?

By 1942, Enrico Fermi had built the first controlled nuclear reactor–Chicago Pile1, the first human-made nuclear reactor–under the University of Chicago's football stands. Using uranium and graphite with control rods, it proved that we could control and sustain fission power, opening the door to the nuclear age.

Within just ten years, tiny reactors powered submarines like the USS Nautilus, which was about the size of a small house and could stay submerged for months. Soon after, commercial nuclear plants came into operation, and scientists began exploring the exciting possibility of using nuclear power in space.

The Soviet Union pioneered early efforts with its RORSAT program from 1967 to 1988. These radar-ocean reconnaissance satellites used compact fission reactors, such as the BES-5 "Buk" design and later TOPAZ systems, to keep ship-tracking radars running for much longer than batteries or solar panels could manage.

Nuclear Batteries for Space

NASA has taken a unique approach to nuclear power in space by mainly using radioisotope thermoelectric generators (RTGs). These are like small, simple "nuclear batteries" about the size of a microwave oven, making them much easier to manage than full reactors.

RTGs don't rely on a chain reaction. Instead, they harness the steady heat produced as radioactive materials–most often plutonium-238–naturally decay. Thermocouples then directly convert the temperature difference between the hot fuel and the coldness of space into electricity. People value RTGs for their simplicity, lack of moving parts, and their ability to provide reliable power for decades, even though the energy output gradually lessens as the fuel wears down. They're especially useful when sunlight is too weak, unpredictable, or blocked by dust.

RTGs have powered some of NASA's most famous space missions. For example, the Voyager 1 and 2 probes, launched in 1977, still send signals from the edge of interstellar space thanks to their RTGs. The Cassini spacecraft explored Saturn for over thirteen years using RTG power, and on Mars, the Curiosity and Perseverance rovers depend on RTGs to keep working through long, dark nights and dust storms—conditions that would otherwise stop solar-powered systems.

On Earth, RTGs are sometimes used in remote or tough environments where plugging into the grid or regular refueling isn't practical—like in isolated navigation beacons or scientific instruments in polar regions. Still, safety concerns, costs, and fuel availability mean they are used mainly for very specialized purposes. In comparison, deep-space missions rely entirely on RTGs because, far from the Sun or in permanent shadow, there's really no better option.

Microreactors

Nuclear microreactors represent the next step forward in space power technology. These small, self-regulating systems can produce much more electricity than traditional RTGs, making them perfect for powering habitats, mining equipment, and manufacturing facilities on the Moon or Mars.

NASA's Kilopower project, which started in 2018, is a fantastic example. A typical Kilopower unit, about the size of a rubbish bin, uses uranium fuel to heat liquid sodium, which then drives a Stirling engine to generate electricity. Just a few of these units could easily power a small lunar or Martian base.

Space microreactors stand out from traditional reactors because they're smaller, simpler, and much safer thanks to passive safety features. Coolants like liquid sodium, molten salt, or helium gas transfer heat effectively in the vacuum of space. Core materials like uranium-molybdenum fuel and reflectors made of graphite or beryllium naturally slow down reactions if temperatures rise, and materials such as boron or cadmium act as "neutron poisons" to automatically dampen the reaction. The reactor's negative temperature coefficient means it reduces power as it heats up and boosts it as it cools down—kind of like a thermostat driven by physics rather than electronics, which is very helpful when human intervention isn't always possible.

Around the world, companies and space agencies are making great progress with these designs. USNC-Tech is developing strong fuels and surface reactors for lunar use, Rolls-Royce is working on compact reactors with advanced cooling systems for lunar missions, and UK-based Perpetual Atomics is creating RTG alternatives using americium-241. There are also exciting concepts for high-power reactors for in-space manufacturing, propulsion systems that could reduce Mars voyages to about one hundred days, and power sources for asteroid mining operations.

Launching Nuclear Reactors

Introducing nuclear systems into space raises unique engineering and regulatory challenges that are both complex and rewarding. The first consideration is mass—every kilogram launched from Earth costs a lot, so designing microreactors that are light and compact, yet still safe with proper shielding and containment, is crucial. This encourages the use of innovative materials and layouts that differ from those on Earth. Safety during launch is a top priority; reactors must withstand accidents like booster failures or uncontrolled re-entry without releasing radioactive materials. To ensure this, extensive analysis and testing are carried out to keep the fuel contained even under severe mechanical and thermal stress.

Regulatory requirements add another layer of complexity. Space nuclear power must comply with national regulations and international treaties. Mission planners need to demonstrate that multiple safety measures are in place and that risks are minimized to very low levels. This involves close coordination among space agencies, nuclear regulators, and international partners.

Designs are also influenced by volume constraints within the launch vehicle fairings. Radiators and other large components are often designed to fold for launch and deploy once in space. Every part must fit within the available space and remain serviceable once in orbit or on the surface.

Despite these challenges, the potential benefits are really exciting. Breakthroughs in radiation-tolerant materials, passive safety features that don't require power or active controls, and modular systems that can be assembled or repaired in space are helping bring a future closer where reliable nuclear power supports long-duration missions far from the Sun.

Impact on Earth

Space-driven nuclear and power technologies are already shaping systems on Earth. The thermal management techniques initially created for reactors and spacecraft have significantly enhanced heating, cooling, and energy recovery systems in buildings and industries, making them more efficient and sustainable.

Safety philosophies and passive safety concepts are guiding the development of next-generation terrestrial reactors, ensuring they are safer for everyone. High-reliability batteries and power electronics developed for lunar and planetary missions are now also playing a vital role in stabilizing electrical grids and supporting the integration of renewable energy sources.

These advances are especially beneficial for remote or underserved areas. Inspired by space technology, compact and sturdy power systems are being developed to provide dependable electricity to off-grid communities, research stations, and emergency facilities—places where diesel generators were once the only option, but now have a brighter, more reliable alternative.

With dozens of firms and research teams pursuing microreactors and advanced radioisotope systems, compact power plants are moving from concept to prototype—capable, in time, of supporting both lunar bases and remote settlements on Earth.

Bottling the Sun

Before we dive into fusion, I wanted to provide a bit of a disclaimer for this very forward-looking idea. Fusion power—harnessing the same process that powers stars—promises almost limitless clean energy for space exploration, but practical fusion reactors for space remain experimental and likely decades away from routine use.

Around 1920, Arthur Eddington first proposed that stars shine by fusing hydrogen into helium. After World War II, scientists set out recreate this stellar furnace process here on Earth. In 1950, Soviet scientists Andrei Sakharov and Igor Tamm created the tokamak, a doughnut-shaped magnetic bottle for hot plasma, while Lyman Spitzer in the US developed the stellarator. Both devices are designed to confine hydrogen plasma heated to over 100 million degrees Celsius, where atomic nuclei can collide and fuse.

By the 1970s and 1980s, large experiments like the Tokamak Fusion Test Reactor in the US, JT-60 in Japan, and JET in Europe made big strides in controlling plasmas, but all they used more power than they generated.

A different approach, inertial confinement fusion, reached a milestone in December 2022 when the National Ignition Facility in California briefly produced more fusion energy from a tiny fuel pellet than the energy delivered by its lasers, even though the entire facility still drew more electricity than it generated; this showed that, in principle, net energy fusion is achievable.

Today, exciting projects like ITER in France and China's upcoming CFETR are working towards demonstrating lasting fusion power by the 2030s and 2040s. In addition, companies such as Commonwealth Fusion Systems and TAE Technologies are exploring smaller, more efficient methods, using high-temperature superconducting magnets, innovative plasma shapes, and alternative fuels.

For space exploration, practical fusion reactors could be game-changers. NASA, DARPA, and other organizations are investigating ideas like fusion-powered propulsion, which could significantly reduce travel times to Mars and beyond, and fusion-powered surface plants that could support lunar colonies during long nights with minimal waste. While these systems are still in the research stage, they could be available in the coming decades.

Even before fusion plants become a reality, their research is already bringing about exciting innovations—ranging from advanced high-field superconducting magnets and precise control systems to new materials capable of withstanding extreme heat and radiation. These innovations also help improve MRI scanners, particle accelerators, and power systems on Earth.

As space exploration fuels breakthroughs in compact power sources, fusion is steadily moving toward practical application. By unlocking the process that powers stars, we're harnessing one of nature's most powerful forces. As one scientist beautifully put it, "We're not just studying the stars anymore. We're learning to build them."

Dark Side of Solar

Solar power seems ideal in space: it offers the beauty of no fuel needs, zero emissions, and an almost endless supply of energy. The main challenge isn't harnessing sunlight, but rather dealing with darkness, dust, and the vast distances involved.

In Earth's orbit, satellites enjoy almost continuous sunlight above the atmosphere, making solar arrays highly effective. They do need to handle brief eclipses and store enough energy for those times, but overall, conditions are rather good.

On the Moon, not so much. Day and night each stretch about fourteen Earth days. During the long lunar night, solar panels produce no power, and temperatures can plummet below -170°C, putting batteries and electronics at risk. Future lunar systems will need to combine solar arrays with advanced storage solutions or backup power sources, along with managing dust that gradually settles on panels and reduces their efficiency.

Mars presents its own set of hurdles. Powerful dust storms can dim sunlight for weeks—a challenge that even ended the Opportunity rover's mission in 2018. Today's Mars missions blend high-efficiency solar panels with improved batteries and smart power management to navigate these events, and some missions use RTGs instead of, or in addition to, solar power.

Beyond Mars, sunlight often becomes too weak for practical solar power in high-energy missions, where nuclear systems often become the better choice.

Since the 1960s, space solar technology has made major strides. Early satellites used basic silicon cells, while today's systems feature multi-junction materials that are far more efficient and resistant to radiation. The International Space Station, for example, uses large, adjustable arrays that rotate to keep facing the Sun.

Innovations like gallium-nitride electronics that withstand extreme temperatures and radiation, along with improved energy storage systems, now enable spacecraft to operate smoothly through long darkness periods. These advances allow solar power to support everything from tiny probes to large space stations.

Interestingly, space-developed solar tech is helping solve energy issues here on Earth, too. Lightweight, high-efficiency panels inspired by satellite tech are now used in portable solar stations for disaster areas, and tough space-grade batteries assist remote clinics in maintaining power during storms.

Techniques initially designed for Mars rovers, like dust cleaning methods, are informing how we protect ground-based solar farms from dust and sand. Smart grid strategies borrowed from spacecraft help detect faults and reroute electricity during outages, while cooling systems crafted for space electronics are extending the life and safety of large batteries on Earth.

Power Beams

Power beaming allows us to send energy wirelessly over long distances using microwaves or lasers. In space, big solar arrays can soak up constant sunlight, turn it into electricity, and send a focused beam of energy to spacecraft or planets.

This idea has been around since the 1960s, when engineer William Brown powered a small helicopter entirely with microwave energy. The process uses radio frequencies that travel through the atmosphere with little loss. On the receiving end, special antennas called "rectennas" catch these microwaves and turn them into usable DC electricity, often achieving efficiencies over 80%. In space, microwave beaming could help satellites share power or reach areas in lunar craters that never see sunlight, where solar panels can't work. NASA and other organizations have run ground tests and small experiments, and several companies are exploring how to use this technology for moon missions or high-altitude aircraft.

Laser power beaming uses highly focused light beams that can stay narrow over long distances, making it ideal for targeting small receivers. On Earth, lasers have successfully kept drones flying and powered remote sensors. In space, similar systems could recharge satellites or support rovers in shadowed regions. Challenges like safety, aiming precision, and atmospheric effects are still being worked out, but ongoing tests are promising.

Looking ahead, some experts believe that by the middle of the century, large orbital platforms could gather solar energy and send it to Earth, supplementing our ground-based renewable energy sources. Whether this becomes cost-effective will depend on launch expenses, regulations, and advances in beaming technology.

Sustainable Power Lessons

Space exploration encourages engineers to view energy as essential for survival, and this perspective is now transforming how we manage power on Earth. For instance, on the International Space Station, waste heat from equipment is captured and reused to keep the cabin comfortable – a concept inspiring projects here on Earth that reclaim heat from subways, data centers, and factories to warm homes and greenhouses. In agriculture, controlled-environment farms adopt space-inspired ideas to make better use of light and heat.

Resilience is another important lesson. Spacecraft need to keep functioning despite faults, so their systems are designed to detect issues early and adapt accordingly. This approach is now guiding smart-grid systems, which can sense outages, isolate faults, and reroute electricity automatically – making energy supply more reliable during storms or equipment hiccups.

Finally, space missions have emphasized the importance of not wasting resources. Closed-loop thinking – recycling air, water, and power – influences industrial parks where waste heat and by-products from one facility become inputs for another. Advanced storage solutions also help balance supply and demand.

The same core principles that help astronauts survive on distant planets – using power efficiently, minimizing waste, and creating systems that can handle failures – are now helping us develop more sustainable and resilient energy systems here at home. Whether supporting a lunar base or a remote village, these practical lessons from spaceflight are increasingly relevant in our daily lives.

Chapter 39

Grabbing Asteroids

Ancient Relics of Our Solar System

Asteroids entered human awareness at the beginning of the nineteenth century, when Giuseppe Piazzi peered through a telescope at the Palermo Astronomical Observatory in Sicily and noticed a wandering point of light moving slowly against the background stars. He named it Ceres, believing for a moment that it might be a comet. Further observation showed that it was neither a comet nor a planet, but a relic from the earliest days of the solar system. Ceres is still the largest object in the asteroid belt and now holds the title of dwarf planet.

These bodies—Ceres and the millions of others—are ancient remnants from the birth of our solar system. Roughly 4.6 billion years ago, when the Sun ignited and planets began taking shape from the swirling disk of gas and dust around it, not all the material successfully coalesced into worlds. Jupiter's gravity shepherded much of the leftover rock and ice into a wide band between Mars and Jupiter, preventing them from merging and preserving them as planetary ingredients frozen in time. Their range is astonishing: from small boulders to mountainous worlds hundreds of kilometers wide. Each one preserves a chemical and geological story older than any rock on Earth.

Humanity's relationship with asteroids is laced with both awe and anxiety. When we see them, we see extinction—craters, shattered continents, fossils and fire. Yet these same perilous wanderers may one day refill our tanks, build our habitats, and give us new materials to thrive in space. Opportunity and danger lie side by side in stone.

Asteroids are packed with water ice, nickel, cobalt, iron, platinum-group metals, carbon compounds, and rare elements that are valuable both in space but also here on Earth. For explorers venturing further out, they seem less like ordinary rocks and more like waypoints in a future solar system supply chain—sources of water for life support and propellant, and metals for building structures, all without the challenge of hauling everything from Earth's gravity well. That said, let's be honest — we're still quite a ways off from making all of this a reality.

As space agencies and companies start to see asteroid mining as a real goal instead of science fiction, the main question has shifted from "Is it possible?" to "How do we do it safely?" Engineers, scientists, policymakers, and legal experts are now discussing the details: how to interact with fragile rubble piles that can shed material at the slightest touch, how to anchor equipment to slowly spinning, low-gravity surfaces, and how to extract resources without disturbing orbits in a crowded, shared environment. Responsible resource extraction is becoming a key part of the conversation.

If done well, asteroid resource could change the game for exploration. Water mined in space can be split into hydrogen and oxygen to make fuel, metals can be used for in-orbit manufacturing, and carbon-rich materials can help support life and provide shielding. Every kilogram we launch from Earth takes energy and costs money; sourcing these materials in space can reduce that load. The real challenge isn't whether these resources are out there — it's figuring out how to access and use them wisely and safely.

The Evolution of Asteroid Exploration

Long before discussions of mining rights or industrial infrastructure, exploration started with simpler questions: What are asteroids made of? How do they hold themselves together? And can a spacecraft approach one without ending its mission?

The first close encounter happened in 1991, when NASA's Galileo spacecraft flew past asteroid Gaspra. Its images showed a battered, irregular world—more fractured and complex than scientists had expected.

Two years later, during its approach to Jupiter, Galileo passed Ida and spotted a tiny companion moon, Dactyl. This discovery changed the field: asteroids could have their own satellites, creating small systems of rock and gravity. Galileo shifted humanity's perspective from distant observation to close-up exploration.

Momentum increased. NASA launched the NEAR Shoemaker mission in 1996, heading toward Eros. After successful orbital studies, engineers tried something new. They guided the spacecraft to land on the asteroid's surface, achieving the first-ever landing. NEAR's survival proved that controlled contact with low-gravity bodies was possible, revealing how these strange worlds behave when touched by machinery.

In 2003, Japan's Hayabusa mission left Earth and visited the asteroid Itokawa. After overcoming multiple technical problems, Hayabusa returned samples to Earth in 2010—making it the first mission to deliver asteroid materials directly to scientists.

Hayabusa2 continued this work by traveling to the carbon-rich asteroid Ryugu. Using a small impactor, it exposed subsurface material and collected pristine samples that arrived on Earth in 2020. Laboratory tests found amino acids and organic molecules, opening new paths in studying life's chemistry.

NASA's OSIRIS-REx launched in 2016 toward the near-Earth asteroid Bennu. When it arrived, the spacecraft found that Bennu was not a solid rock but a loose "rubble pile" held together by gravity. Instead of trying to land fully, OSIRIS-REx performed a quick touch-and-go maneuver, collecting material with high precision. It returned the largest-ever asteroid sample in 2023. The mission proved that even delicate aggregates can be explored safely with careful engineering.

Before anyone thought of mining asteroids for riches, explorers first had to solve a much older mystery: what were these wandering rocks made of, and could a tiny spacecraft even survive meeting one face-to-face? Instead of drills or dynamite, the earliest missions began with a mix of asking a ton of questions and staring through some rather powerful telescopes.

Lessons from Space Missions

This has led to a new generation of technologies designed to respond to local conditions. Some landers use articulated legs that adjust to uneven terrain, spreading their weight with small actuators to prevent sudden collapse. Anchoring systems rely on soft capture harpoons or low force drills that grip the surface without cracking it. For metallic asteroids, electromagnetic nets could collect loose particles without digging at all.

All of this must happen autonomously. Communication delays prevent real-time joystick control of a lander. For many near-Earth asteroids, the delay is tens of seconds one way; for main-belt targets, light time often reaches tens of minutes. Real missions like OSIRISREx operated with an eighteen to twenty-minute delay each way while working around Bennu. A command such as "stop" would arrive long after the spacecraft had already acted.

Asteroid landers, therefore, will need to sense hazards, choose footholds, and manage their descents in real time. AI-driven onboard systems would fuse data from cameras, lidar, and accelerometers to map drifting terrain, spot dust plumes and shifting rubble, and select safe trajectories without waiting for instructions from Earth. These autonomous controllers would constantly adjust thrust and leg positions, tune how hard a drill or sampling head presses into the surface, and decide when to retreat—so that every step, hop, or arm movement protects both the spacecraft and the fragile ground beneath it. In the long term, the same capabilities could guide robotic refueling depots, in space factories, and even construction crews assembling habitats far beyond Earth.

These missions are teaching us how to interact with the solar system in ways that preserve ancient worlds, maximize safety, and extend exploration into regions once considered unreachable. As AI and autonomy mature, robots will increasingly become onsite partners rather than remote puppets, learning from each contact with an asteroid and helping to write the operating manual for the first generation of offworld miners and builders.

Building an Asteroid-Based Economy

Early estimates of asteroid value generated sensational headlines. Some calculations suggested that a single metal-rich asteroid contained more nickel and iron than all of Earth's historical production. Economists quickly pointed out that returning large amounts of platinum group metals to Earth would cause prices to collapse. The real value lies in using resources close to where they are found. Water can be turned into propellant in orbit. Metals become feedstock for 3D-printed space structures. Carbon-rich materials feed life-support systems. Asteroids enable local manufacturing, cutting the need to launch heavy materials from Earth's deep gravity well.

Modern companies are advancing the required technologies. TransAstra is pursuing solar thermal mining, using concentrated sunlight to extract materials without heavy drilling. AstroForge is testing metal processing techniques in microgravity. These approaches aim to gather resources while minimizing forces on fragile structures. Meanwhile, missions such as Hayabusa2 and OSIRIS-REx provide detailed maps of composition and mechanical behavior—the data needed to design mining systems grounded in reality rather than speculation.

Even so, the full economics of developing, launching, extracting, and using asteroid resources remain unproven. While that work continues, many of the underlying technologies already find use in extreme environments on Earth—from deep ocean ridges to remote polar regions—where fragile geology, limited access, and high operational risk mirror asteroid conditions. As these methods mature, they create safer, cleaner, and more efficient ways to explore difficult terrain, proving their value long before the first commercial asteroid is touched.

The Bigger Picture

Asteroid exploration and extraction technologies ripple far beyond the space sector. Tools created for navigating low gravity landscapes now assist autonomous robots in the deep ocean. Mineral sorting systems developed for asteroid samples help improve critical mineral processing on Earth. AI techniques for tracking tumbling asteroids contribute to safer underground mining. Legal frameworks originally drafted for shared space resources inform more sustainable approaches to resource use at home.

The value of asteroid work unfolds on three levels: direct use of resources in space, lower launch costs by sourcing materials offworld, and the transfer of innovation back to Earth. By treating these ancient bodies with care, humanity gains new tools to support exploration, promote sustainable technology, and expand its reach. The story of asteroids shifts from fear to opportunity—from scattered debris to a foundation for a broader human presence in the solar system.

Chapter 40

Anatomy of a Space Probe

Our Brave Cosmic Ambassadors

Long before humans set foot on the Moon, we sent robotic probes venturing into the vast, chilly emptiness of space, exploring alien worlds we'd never see firsthand. From the fiery volcanoes of Venus and the ancient riverbeds of Mars to the icy secrets of outer planets, these little explorers have been our brave cosmic ambassadors, traveling through harsh environments—furnace-like atmospheres, intense radiation, and the endless midnight cold between planets. Crafted from durable, lightweight materials and equipped with cutting-edge instruments, they turned tiny points of light into detailed worlds. Some, like the remarkable Voyagers, keep journeying beyond the solar system even now, carrying our story into the unknown—calling back signals that reach us across the vast emptiness, waiting for whoever or whatever might be listening.

In this chapter, we take a closer look at the ingenious technologies that make space exploration possible—things like propulsion, communication, shielding, and most importantly, the tools that unlock the secrets of the universe. Step by step, we follow the missions that charted the unknown, discovered new realms, and expanded our understanding of the cosmos.

When you think of a space probe as a small mechanical lifeform; its design makes more sense. It has a skeleton, eyes, an inner ear, a nervous system, and a brainstem – all working together to keep it alive, balanced, and capable of functioning independently from so far away from home.

Let's explore how these little spacecraft that are millions or billions of miles away are still faithfully calling home and communicating more reliably than your teenager.

Wee Instruments

Think of a probe's instruments as its senses. When exploring the expanse of space, we need tiny but mighty "sense organs" because spacecraft have strict limits on how much they can weigh and how much power they can use. Missions like New Horizons, Juno, and Europa Clipper exemplify this with their compact but powerful scientific tools.

These missions depend on cutting-edge materials and technology. For example, carbon fiber reinforced polymers cut down instrument weight by more than half compared to traditional aluminum, all while keeping the structure strong—like in Europa Clipper's REASON radar with its lightweight, intricately-shaped antenna. Special ceramic parts made from silicon carbide and alumina stay stable across extreme temperatures from -200°C to +200°C, yet weigh a lot less than metal ones. This is crucial for instruments like Europa Clipper's camera system, which must remain perfectly aligned despite thermal changes.

Even more impressive, 3D-printed titanium allows for complex instrument housings and structures that reduce weight by a quarter or more, creating shapes impossible with traditional methods.

The benefits of these space innovations go way beyond space itself. Medical devices, for instance, have seen huge improvements thanks to sensors inspired by spacecraft tech—these tiny, non-invasive diagnostic tools have made health monitoring easier and more accurate. MEMS (microelectromechanical systems) accelerometers used in Europa Clipper's gravity tests now help run implantable heart monitors smaller than a grain of rice. Consumer gadgets have also advanced; GaN-based power electronics, originally developed for space, now make fast charging for smartphones and laptops a reality.

The miniaturization techniques perfected for space missions have democratized satellite technology, making it possible to launch affordable CubeSats the size of a shoebox, with capabilities that once needed large, costly spacecraft. Universities and developing countries can deploy Earth-observation tools at a fraction of the earlier costs, thanks to sensor tech inherited from billion-dollar flagship missions. Even health wearables have been transformed by space innovation, with ultra-thin, flexible sensors that can continuously track vital signs without being noticeable to the wearer.

All these examples show how the tough challenges of exploring deep space push us to develop new technologies—many of which end up improving health, connectivity, and understanding back on Earth.

Spin Class

Once a spacecraft reaches space, it needs to monitor its orientation and ensure it is pointed in the right direction. This is a bit like the probe having an inner ear to give it a sense of balance.

One of the simplest methods engineers use for this is called *spin stabilization*. Remember Newton's First Law of Motion—objects in motion stay in motion unless something causes them to stop or change direction? Well, spin stabilization is basically the same idea, but for spinning objects.

When you're in a car and the driver brakes suddenly, your body wants to keep moving forward, much to the annoyance of your seatbelt. This is inertia, which resists changes in movement. For spinning objects, a similar property called *angular momentum* tries to keep the object spinning. You see this with spinning tops that stay upright and bike wheels that help bicycles stay balanced.

Sometimes, things can get tricky. In 1985, cosmonaut Vladimir Dzhanibekov was spinning a wingnut that initially turned smoothly, then suddenly flipped over and spun and wobbled unpredictably. This phenomenon became known as the Dzhanibekov Effect, or in physics terms, the *Intermediate Axis Theorem*. It says that if you try to spin an object around its middle axis—that is, neither the longest nor the shortest—even a tiny push can cause it to flip or tumble unpredictably.

You can observe similar behavior on Earth too. A gymnast can spin along their long axis, like a pirouette, or along their short axis, like a somersault. But if they spin around their middle axis, like doing a sideways cartwheel, controlling that motion becomes much more difficult. Or think of a football (American football, that is!) thrown without a perfect spiral—it wobbles in an unstable way, making its motion unpredictable. In each case, the object is spinning around its least stable axis.

Because of this, spacecraft are designed to spin around their most stable axis—typically the one with the largest or smallest moment of intertia —to ensure smooth, predictable rotation. Spinning in any other way, especially in the vacuum of space where there's nothing to slow it down, can lead to problems.

I'm sure you've seen a figure skater spin faster as they pull their arms closer to their body, and slow down as they stretch their arms out. The total angular momentum remains the same (it's conserved!) but moving their arms in or out changes their spin speed.

Spacecraft use these principles in a couple of ways. Spin stabilization helps keep a probe balanced and steady without needing constant adjustments. Some satellites use reaction wheels—weighted wheels inside that can spin faster or slower—to control the spacecraft's orientation. This method is very reliable and simple, allowing the satellite to stay balanced as it travels through the emptiness of space.

The Evolution of Spin

In its simplest form, spin stabilization means spinning the entire spacecraft. Picture Pioneer 10 and 11, launched in the 1970s, calmly twirling as they sped toward Jupiter and beyond. This steady spin made them surprisingly stable platforms and saved precious fuel on their long journeys.

As missions grew more complex, engineers got creative with dual-spin designs. The Galileo spacecraft, for example, used a spinning section for stability while another section stayed still, keeping sensitive instruments and antennas locked onto their targets. It's like our figure skater holding her head still while her entire body spins below her—admittedly, she will quickly turn her head as well, but you get the point!

These days, spacecraft like Juno use a hybrid approach: they spin at about two revolutions per minute while in orbit around Jupiter. This way, sunlight is spread evenly over the solar panels, and instruments sweep across Jupiter's clouds in a smooth, efficient rhythm.

Spin stabilization's biggest advantage is simplicity. Fewer moving parts mean there's less to break, and you don't need lots of computers or constant thruster bursts to keep things pointing the right way—key for spacecraft designed to operate for decades without repairs.

That said, spinning everything isn't perfect. If you want cameras aiming in one direction and antennas in another, pure spin makes things awkward. Engineers solve this with "despun platforms"—special sections that stay still as the rest of the craft spins, keeping instruments or dishes locked onto their targets.

Eventually, technology advanced, and engineers moved to what's called the *three-axis stabilization*. Instead of spinning, the spacecraft balances itself along three directions—up-down, left-right, and front-back—so any side can point at any target. Missions like Voyager, Cassini, New Horizons, and Europa Clipper all use this technique, relying on little thruster puffs or spinning reaction wheels inside to nudge them in just the right direction.

You can see the same physics at play all around us. Earlier generations of communication satellites used spin stabilization to stay pointed, gyroscopic principles keep bike wheels upright, and even your smartphone's camera uses related ideas for shake-free photos.

The lessons learned designing all these stabilization systems improved not only space engineering but also control systems, materials science, and mechanical design right here on Earth.

So whether it's a pure spinner like Pioneer, a split-spin design like Galileo, or a hybrid like Juno, spin stabilization is a brilliant example of physics in action—helping fragile machines stay steady on the longest, loneliest journeys humanity can send them.

Precision Navigation

Navigating deep space isn't just about aiming for distant worlds—it's about hitting targets smaller than a pinhead, billions of miles away. A probe's navigation system is its mind at work—interpreting signals, predicting trajectories, and figuring out how to hit a target billions of kilometers away.

Deep-space navigation came of age with missions such as the Pioneer probes. Imagine engineers tracking a probe that's billions of miles away from Earth, trying to figure out its position and speed. They use tiny frequency shifts in radio signals, called Doppler tracking, which can detect even the smallest changes in the probe's velocity, sometimes as small as a millimeter per second—the same speed at which an ant crawls! This gives mission controllers an incredible ability to know exactly where the probe is, no matter how far it's traveled.

Voyager went further by "catching a ride" on gravity itself. By passing by planets like Jupiter, Voyager was able to pick up speed—over 35,000 miles an hour—without using any extra fuel. These gravity assists work like cosmic slingshots we talked about before, and planning them takes careful calculations of orbital mechanics made years ahead of time. Without these maneuvers, reaching the far corners of our solar system wouldn't be possible.

Future missions like Europa Clipper are gearing up for even more challenging adventures. They'll use increasingly autonomous navigation, so the spacecraft can adjust its course on its own without waiting for every instruction from Earth. Equipped with onboard computers designed to withstand Jupiter's harsh radiation, they can make quick decisions while performing complex calculations that consider the gravitational pull of nearby moons and planets.

It's amazing how the advanced navigation techniques from space exploration have become such an integral part of our everyday lives here on Earth. Every time you use GPS, your phone is using the same fundamental physics—like precise timing, Doppler shifts, and orbit prediction—that scientists originally developed to guide spacecraft across the solar system. The everyday routes your car takes, the paths ships follow, and the way planes and air traffic controllers keep us safe—all hinge on error correction and prediction methods that were once only used in deep space missions. How incredible is it that the same level of precision used to explore our solar system now helps us navigate confidently and accurately, whether we're discovering a new city or sending robots on interstellar journeys?

Radar Mapping

As we saw before, Synthetic Aperture Radar (SAR) helps us see through clouds and darkness on Earth—this is the vision and "x-ray" for satellites and probes. But this incredible technology has been just as revolutionary for exploring distant worlds where traditional cameras simply will not work.

Below thick, poisonous clouds, there are mysterious landscapes, or oceans that flow beneath icy crusts of moons—places we might never be able to see without SAR technology. By bouncing radio waves off surfaces and measuring what comes back, spacecraft can "see" what would otherwise stay hidden from us.

Deep Space SAR Pioneers

Missions like the Soviet Venera Landers, NASA's Cassini-Huygens, and the upcoming Europa Clipper have all used special radar technology to see what our eyes cannot. Back in the early 1980s, Soviet Venera missions were the first to use this technology to map Venus. They revealed an amazing landscape of mountains, valleys, and lava flows that would otherwise remain hidden under Venus's thick clouds. Without this radar technology, we would have no idea what Venus actually looks like.

Years later, NASA's Cassini spacecraft used similar radar to peer through the thick orange haze surrounding Saturn's moon Titan. The radar showed us a world with lakes and seas (not of water, but liquid methane), river channels, and even dune fields that look surprisingly similar to Earth despite being made of completely different materials.

In the near future, the Europa Clipper mission will take this further to penetrate ice to look beneath the frozen surface of Jupiter's moon Europa. This will help scientists understand how thick the ice is and what it's made of, potentially giving us our first glimpse of the ocean scientists believe exists underneath.

The same radar technology developed for these space missions has changed how we study our own planet too. Today, radar satellites help us track deforestation, measure sea ice thickness, monitor natural disasters in real time, and detect tiny movements in buildings and bridges before failure—potentially saving lives.

Digital Cameras

Back in the summer of 1979, people everywhere watched in awe as grainy images flickered onto their TV screens—towering clouds swirling across Jupiter's atmosphere, like cream in coffee and larger than Earth itself. Suddenly, alien planets were dynamic and dramatic. All of this came from the Voyager spacecraft, and hidden inside was a new technology with a long future ahead: the Charge-Coupled Device, or CCD—a tiny chip that changed how we see the universe, and ourselves. If SAR is the deep vision of a probe, CCD is its daylight vision: sharp, sensitive eyes that let us use our own eyes through the probe's eyes.

At first glance, a CCD is just a silicon chip carved into tiny squares called *pixels*. But each pixel is a small sensor, waiting to catch a photon of light and produce an electronic signal based on how bright or dim the light is. Linking up millions of these signals creates a vivid, digital picture (a picture of digits, if you will).

The CCD's story begins at Bell Labs in 1969. While working on computer memory, inventors Willard Boyle (1924-2011) and George E. Smith (1930-2025) accidentally discovered how a slice of silicon could "see" light. Stack enough pixels together, and with each flash of light, an image appears. They had no idea their tinkering would eventually win them a Nobel Prize—or transform exploration itself.

For space science, CCDs arrived at exactly the right time. Before then, probes often relied on film cameras—requiring actual film canisters shot in space and sent back to Earth, a system completely useless for deep space. Some spacecraft used tube-like TV cameras, which struggled with hazy, dim worlds. Sending high-tech "eyes" beyond Earth was a challenge—until CCDs made the leap possible.

Voyager was the first to break the mold, launching in 1977 with CCD cameras prepared for the great unknown. Suddenly, crisp images of Jupiter's cloud belts, Saturn's shimmering rings, Uranus's pale turquoise glow, and Neptune's icy heart poured in. We could finally see these places for real–not just as artists' sketches. Later missions, like Galileo, at Jupiter and Cassini at Saturn, used even sharper CCDs, capturing volcanic eruptions on Io, hidden oceans inside icy moons, and details no one had glimpsed before.

What made CCDs special is how amazingly sensitive they are–they can detect the faintest cosmic signals that would be just darkness on film cameras. This means scientists can get instant digital images and analyze mysterious features immediately, rather than waiting months. Plus, these chips can be adjusted to see ultraviolet, infrared, and X-rays, uncovering hidden secrets that our eyes can't see.

The magic of CCDs didn't stay in space. It quickly spread into telescopes, microscopes, hospitals, and, finally, our everyday devices. When you get a high-tech medical scan, snap a picture on your phone, or check out with groceries, a CCD is working behind the scenes. Our smartphone and digital camera revolution can be traced directly back to the same invention that brought Saturn's moons and Pluto's icy plains into our living rooms.

From an inspired lab experiment, to a chip soaring past Neptune, to the phones in our pockets, the journey of the humble CCD reminds us: every great leap for science can end up changing daily life in ways no one expected. These chips may not be household names, but they've given us new eyes–allowing us to explore distant worlds and the everyday wonders around us, in astonishing clarity.

Autonomous Operations

Picture yourself driving a winding, foggy mountain road. Every turn, you have to call someone miles away for instructions and then wait ages for their response before you can steer or slow down. It would be a nightmare–unsafe, slow, and nearly impossible. That's exactly how deep space exploration would work if spacecraft couldn't operate on their own. This is the space probe's brainstem, handling emergencies, adjusting to hazards, and generally keeping it alive.

That's where autonomous systems come in. They give spacecraft the intelligence to think and act independently, thousands–even billions–of kilometers from Earth. With radio signals crawling through space, delays can be minutes or even hours. There's no way for mission control to guide every move in real time, so probes need onboard smarts–systems that let them sense problems, make decisions, and solve surprises all by themselves.

Consider a probe exploring a comet or moon. Dust clouds might appear out of nowhere, or a thermometer might break. With no one nearby to help, the spacecraft must quickly figure out what's happened and take action before disaster strikes. Sensors, smart software, and algorithms allow these robots to understand their situation, diagnose issues, and run critical scientific experiments–all without human intervention.

Without this autonomy, missions would spend most of their time waiting for instructions–and valuable discoveries might slip away while Earthbound scientists scramble to catch up. But with the ability to "think for themselves," probes go farther, adapt faster, and capture astonishing science from places no person could reach.

Take Galileo, for example. As it orbited Jupiter in the 1990s and early 2000s, Galileo faced numerous hazards–from radiation overloads to hardware glitches billions of miles from home. Its backup systems automatically activated when things went wrong: it could swap out failed parts, reposition its antenna, and continue sending back data from the stormy Jovian moons even while operating independently.

Voyager 1 and 2–launched in 1977 and now humanity's most distant explorers–are also masters of autonomy. With radio signals taking more than 20 hours one way, they operate on pre-programmed instructions: steering by the stars, adjusting their power, and operating delicate instruments with minimal help from Earth. That self-reliance made their legendary "Grand Tour" of the outer planets possible and keeps them communicating with us from interstellar space today.

New Horizons, zipping past Pluto into the Kuiper Belt, takes autonomy even further. It protects itself from hazards, steers itself using star trackers, and captures images of distant worlds with little human oversight– crucial when the entire flyby of a new world happens faster than messages can travel from Earth.

The next step is Europa Clipper, which will orbit Jupiter's icy moon starting in the mid-2020s. It will contend with intense radiation and tricky orbits by using its sensors and software to protect sensitive instruments and gather the best possible science data on each flyby.

Like many other things, these innovations don't just stay in space. Autonomous technology from deep space now powers systems on Earth: GPS and navigation systems, smarter drones and cameras, even medical monitors that keep tabs on patients without constant doctor oversight. Algorithms originally designed to help spacecraft avoid danger have made today's self-driving cars and home robots safer and more reliable.

Rugged electronics built for space–designed to withstand harsh radiation, vacuum, and temperature extremes–have also led to more reliable satellites for weather, communications, and Earth observation. They provide vital data for farmers, disaster response teams, and climate scientists, keeping us better informed and better prepared.

Space robots are intricate feats of engineering, each part of their anatomy performing a specific role honed over decades of research in laboratories around the world. Their anatomy reflects our own: perception, balance, memory, and decision-making. As they learn to survive on their own, they help their makers build smarter and safer technologies back on Earth.

Chapter 41

Giant Ball of Fire

The Sun and Life

Many people picture the Sun as a giant, glowing ball of gas, sitting at the center of our solar system. In reality, it's much more intense—a raging sphere of plasma, seething with super-hot, electrically charged particles (mostly hydrogen and helium). Deep inside, hydrogen atoms smash together to create helium, releasing massive blasts of energy. This energy becomes the light and warmth that radiates out into the solar system, fueling nearly every natural process on Earth.

What would happen if the Sun suddenly switched off? Even though our planet has some heat deep inside, it would freeze and darken; life as we know it would be impossible. Our oceans and air would chill without the power of sunlight. The Sun lifts water to form clouds and rain, lets plants make food, and whips winds and weather around the globe. Without sunlight, our food chains would not exist and most living things—and rainbows or the northern and southern lights—just wouldn't exist.

Rainbows paint the sky after rain when sunlight bends through tiny droplets, splitting white light into every color—red, orange, yellow, green, blue, indigo, and violet—in a familiar arc stretching across the clouds.

The northern lights, or *aurora borealis*, and their southern twin, the *aurora australis*, put on a show of their own. Streams of charged solar particles streak toward Earth, riding the solar wind. When they crash into our planet's magnetic shield and upper atmosphere, they smash into oxygen and nitrogen atoms high above the poles, releasing brilliant flashes—green and red ribbons from oxygen, blues and purples from nitrogen—dancing across the sky.

Rainbows and auroras are living reminders that the Sun connects to every single breath we take, every bit of color and spectacle we see and enjoy on Earth.

Eruptus Maximus

Even with centuries of observation, some of the Sun's most important behaviors remain difficult to predict: its solar wind, magnetic field, and bursts of radiation.

A constant stream of charged particles, the solar wind, flows outward from the Sun and shapes planetary environments. Its tangled magnetic field creates sunspots, solar flares, and coronal mass ejections that occasionally send more intense bursts toward us, able to disturb satellites, radio links, GPS, and even power grids. Earth's own magnetic field and atmosphere deflect and absorb much of this radiation, which is one reason life at the surface can exist at all.

Scientists study these processes to forecast "space weather," protect technology, and test fundamental physics. Solar research benefits Earth through improved satellite communications, GPS navigation, power grid protection, and technologies like solar panels and heat-resistant materials. Understanding solar behavior helps guard against potentially costly storms.

To understand these forces in detail, scientists have launched a series of daring missions that fly closer to the Sun than anyone once thought possible.

Chasing the Sun

In the 1970s, the joint German-American Helios missions became the first to approach the Sun closely—within twenty-seven million miles. They studied solar wind and magnetic field interactions using instruments to measure wind speed, density, temperature, and magnetic strength. Though designed to last for several years,

Helios 2 operated for fourteen years, discovering how solar wind speed varies with distance and detecting micrometeoroids near the Sun. Helios paved the way for missions that would venture even farther into the Sun's domain.

Launched in 1990 by NASA and ESA, Ulysses took an unprecedented trajectory over the Sun's poles by swinging past Jupiter. This gave scientists their first view of polar regions previously impossible to study. The mission revealed faster solar wind near the poles and a more complex magnetic field than expected. It also showed how cosmic rays access the inner solar system through the poles, helping scientists understand how the Sun's magnetic field shields Earth.

NASA's Parker Solar Probe, launched in 2018, is flying closer to the Sun than any spacecraft in history - approaching to about six million kilometers from its surface, several times closer than Helios 2. Using repeated fly-bys of Venus, it builds up enough speed to reach around 700,000 km/h while its instruments hide in the shadow of a 4.5-inch carbon-composite heat shield that faces temperatures of about 1,400°C.

Even in its early orbits, Parker has made important discoveries. It has traced much of the solar wind back to "coronal holes"—regions where magnetic field lines open into space—and found strange "switchbacks" in the magnetic field, sudden reversals in direction that no one expected to see so close to the Sun. Findings like these are reshaping how we think the solar wind forms and helping us better predict storms that can affect satellites and power grids on Earth.

While spacecraft race toward the Sun, some of our most powerful tools remain here on Earth, ground observatories complement spacecraft missions. In Hawaii, the Daniel K. Inouye Solar Telescope observes features as small as eighteen miles across from ninety-three million miles away, studying sunspots and solar flares. Radio telescopes monitor solar activity by detecting radio waves from flares and eruptions that may affect Earth's technology. Coronagraphs block the Sun's bright disk to reveal its fainter corona, where solar wind originates and coronal mass ejections occur.

What We've Learned

From heat-tolerant spacecraft like Helios, Ulysses, and Parker Solar Probe to a global network of observatories on the ground, scientists are stitching together a more complete portrait of our star. Each mission focuses on a different mystery: the fusion reactions deep in the core, the shifting surface of sunspots and flares, and the wispy, turbulent corona where the solar wind escapes into space.

What we learn goes far beyond satisfying scientific curiosity. Understanding the Sun's rhythms helps us build better satellites, safeguard power grids, and predict storms before they interfere with communications or disrupt aircraft. Research on our star also shapes everyday technologies, from solar panels and forecasting models to the systems that keep global communication networks running. As we unravel how the Sun works, we learn how to better protect climate-sensitive infrastructure and the technologies that connect an increasingly interdependent world.

Chapter 42

Mysterious Contradiction

Mercury's Wee Mysteries

Mercury, the smallest and swiftest planet in our solar system, orbits closest to the Sun, yet it remains one of our most mysterious neighbors. Despite being practically next door in cosmic terms, only two spacecraft have ever visited this extreme world, providing limited yet fascinating glimpses into its nature and history.

The first to get a close-up look was Mariner 10, launched by NASA in 1973. Trailblazing in more ways than one, Mariner 10 became the first craft ever to visit Mercury and the first to use a flyby of Venus to reach its target. During three flybys in 1974 and 1975, the probe snapped striking images of about forty-five percent of Mercury's battered, cratered crust—a landscape that looked uncannily like our Moon.

Mariner 10's instruments made some incredible finds for such a small, seemingly plain planet. It detected an incredibly thin layer of atmosphere, or exosphere, made of lonely atoms floating above the surface. Mercury also turned out to have an unexpectedly strong magnetic field, and readings hinted at a vast, dense iron core below—a major clue to Mercury's origins. Though Mariner 10's mission was shorter than a year, its discoveries set the stage for all future missions to this little planet in the Sun's glare.

MESSENGER

More than thirty years passed before NASA returned to Mercury. In 2004, the MESSENGER mission set out to do what Mariner 10 could not: stay. Instead of a few quick flybys, MESSENGER would orbit the planet, mapping and studying it in unprecedented detail. After a seven-year journey, the spacecraft entered Mercury's orbit in 2011 and spent over four years circling Sun's innermost planet before its planned descent in 2015.

Equipped with an advanced suite of instruments—cameras, spectrometers, magnetometers, and altimeters—MESSENGER explored everything from Mercury's surface and magnetic field to its thin exosphere, uncovering a world far more dynamic than scientists ever expected.

MESSENGER's achievements have changed our understanding of Mercury. It detected clear evidence of ancient volcanic activity that had resurfaced large parts of the planet, reshaping its landscape. The mission also discovered the surprising presence of water ice trapped in permanently shadowed craters near Mercury's poles, even though the planet is scorching hot. MESSENGER also revealed that Mercury's diameter has shrunk by about fourteen kilometers (nearly nine miles) as its iron core has cooled and contracted—a captivating hint about the planet's geological story. On top of that, its elemental analyses showed unexpected levels of volatile elements like potassium and sulfur, which challenge previous ideas about Mercury's formation and suggest it has a more intricate history than we had thought.

The knowledge gained from Mariner 10 and MESSENGER has had wide-reaching impacts beyond planetary science. Engineering innovations stemming from the need to survive Mercury's extreme environment—where temperatures can swing dramatically from 800°F (427°C) in sunlight to -290°F (-179°C) in shadow—have led to advances in heat-resistant materials and thermal management systems now used in a variety of industries on Earth. Studies of Mercury's magnetic field have enhanced our broader understanding of how planetary magnetic fields originate and protect planets, informing research about Earth's magnetosphere and space weather effects. The imaging and mapping techniques developed for these missions have also contributed to improvements in remote sensing technologies used to monitor Earth's surface, resources, and environment.

Together, Mariner 10 and MESSENGER have turned Mercury from one of the least-known worlds into one of the most intriguing—showing that even the smallest, harshest places in our solar system can change the way we see planets, and ourselves.

Chapter 43

Earth's Toxic Sister

Our Neighbor Venus

Venus, often called Earth's sister planet because of its similar size and proximity, has long captured the imagination of scientists and explorers alike. Although it's our closest neighboring planet, Venus presents an array of extreme and hostile conditions that make exploration both exciting and challenging. Its thick atmosphere is packed with dense, toxic clouds of sulfuric acid, and surface temperatures can climb to around 900°F (475°C)–hot enough to melt lead. The pressure at its surface is nearly 100 times greater than Earth's, creating a punishing environment for spacecraft and instruments.

Thanks to these hurdles, missions to Venus have been incredible feats of engineering, transforming our understanding of this enigmatic planet. Far from being a temperate Earth-like twin, Venus revealed itself as a world of extremes–shaped by volcanic activity, crushing pressure, and a dense, toxic atmosphere driven by a runaway greenhouse effect. The subsequent missions–even from different parts of the world–have gradually uncovered Venus's veil one layer at a time, revealing not a paradise but a furnace.

The Pioneers: First Encounters with Venus

The Soviet Union's Venera program pioneered Venus exploration. In 1961, Venera 1 was lost before arrival but provided valuable experience. In 1965, Venera 3 became the first humanmade object to enter another planet's atmosphere, even though communication was lost before it could report back.

In 1967, Venera 4 made the first direct measurements of Venus's atmosphere, revealing that it was mainly carbon dioxide with nitrogen and sulfur compounds—evidence that explained the planet's extreme heat. Then in 1970, Venera 7 achieved the first successful soft landing on another planet's surface, confirming temperatures of about 465°C and pressures ninety times higher than Earth's.

Later Venera missions advanced our knowledge even further. In 1976, Venera 9 and 10 transmitted the first surface images, showing rocky terrain under a dim orange sky. Venera 13 and 14 sent back color photographs revealing reddish-brown volcanic soil. Venera 15 and 16 used radar to map surface features hidden beneath the clouds, while the Vega missions in 1985 deployed atmospheric balloons that rode through super-rotating winds high in the atmosphere.

On the American side, Mariner 2 holds a special place in the history of space exploration as the first successful mission to another planet. Launched by NASA in August 1962, it flew within about 21,600 miles (34,800 kilometers) close to Venus. With its microwave and infrared radiometers, magnetometers, and charged-particle detectors, it revealed that Venus's surface temperatures soar above 800°F (427°C). It also uncovered that the planet is covered in a thick, hot atmosphere instead of oceans. Its success proved that spacecraft can successfully travel to and explore distant worlds, opening the door for all the incredible planetary missions that followed.

Unveiling the Hidden World

In 1989, NASA's Magellan used radar to explore Venus and maps its surface through the thick cloud cover. Between 1990 and 1994, over more than 8,000 orbits, it created detailed 3D maps of volcanic plains, mountains, craters, and tectonic features. This confirmed how volcanic activity significantly shapes the planet's surface. Magellan showed just how amazing radar mapping can be for exploring planets and helped push forward advances in space imaging technology. Its findings hinted at recent volcanic activity and revealed that Venus's surface has been reshaped by volcanic and tectonic activity, some of it possibly recently. Even in such a tough environment these insights later helped improve remote-sensing techniques that we now use to study Earth.

Years later, in the Venus Express mission, launched by the ESA in 2005, scientists shifted their focus from the surface of Venus to its atmosphere. They sent spectrometers to analyze the atmospheric chemistry, cameras to capture images of clouds and surface features, and various instruments to explore temperature and circulation patterns. The Venus Express mission revealed complex circulation in the planet's atmosphere, like super-rotating winds that race around the globe much faster than Venus itself spins—building on early hints seen by the Vega balloons. It also deepened our understanding of cloud chemistry and the role of sulfur compounds, found evidence of lightning, and spotted temporary ultraviolet dark spots that might be linked to volcanic activity. Even without landing on the planet, Venus Express greatly expanded our knowledge of Venus's weather, climate, and atmospheric structure.

Back in 2010, Japan's Akatsuki joined the quest. After a rocky start with an initial orbit insertion failure caused by a valve issue, the engineers showed incredible resourcefulness—using smaller thrusters, they managed to place the spacecraft into Venus orbit in 2015 after spending five years orbiting the Sun. A truly remarkable comeback. Akatsuki is now busy studying cloud movements, temperatures, and weather phenomena in various wavelengths, observing everything from daytime cloud patterns to nighttime emissions. Although by the mid-2020s, the mission faced some communication and attitude-control challenges, its extensive history has offered us a valuable, detailed look at Venusian weather over time.

What We've Learned

So far, only the Soviet Venera landers have braved the surface of Venus, but each international mission has contributed new insights to our understanding. Today, we see Venus as a toxic world of extremes: a planet cloaked in acidic clouds, shaken by volcanic activity, with an atmosphere 90 times thicker than Earth's, swept by hurricane-force winds. At the same time, Venus acts as a natural laboratory for studying the greenhouse effect, illustrating how a once Earth-like planet can become uninhabitable when heat gets trapped for ages—offering important lessons for our own climate.

The tough challenges faced by Venus missions have led to advances in spacecraft design, heat-resistant materials, and thermal management systems, many of which are now used on Earth in high-temperature industries and tough environments. As scientists continue exploring Venus's geology and chaotic weather, big questions remain: Was it ever more Earth-like? Could it have held oceans or life? Or will it always stand as our "toxic twin," warning us how a familiar world can go terribly wrong?

Every Venus mission so far shows human perseverance and ingenuity—a story of brave machines, smart science, and explorers eager to uncover the secrets of our dangerous neighbor.

Chapter 44

Gas Giant

Stormy Jupiter

Jupiter dominates our solar system as the heavyweight champion—by far its largest and most spectacular planet. Known as a "gas giant," Jupiter is mainly composed of hydrogen and helium, unlike Earth's rocky surface. Its diameter alone—about 143,000 kilometers (nearly 89,000 miles)—exceeds that of Earth by more than eleven times. If you could hollow out Jupiter, you could fit over 1,300 Earths inside. Nevertheless, it is not particularly heavy for its size, since it is made mostly of lightweight gases.

Beneath Jupiter's swirling, striped clouds, scientists suspect there is a dense core surrounded by strange metallic hydrogen—a form of compressed hydrogen that behaves like an electrically conductive fluid and may help generate Jupiter's powerful magnetic field, unlike any other. Jupiter spins incredibly fast, completing a full rotation in just ten hours, causing its middle to bulge outward like a spinning water balloon.

Take a closer look, and you will see Jupiter's atmosphere arranged in bold bands that weave around each other in opposite directions—a hallmark of turbulent, mighty winds. The Great Red Spot continues as a gigantic cyclone, larger than Earth and visible for centuries. Other clouds swirl with ammonia ice, sulfur, and phosphorus, while thin dusty rings and faint auroras flicker above its poles.

Jupiter has always fascinated scientists. Over the years, robotic explorers have studied its wild weather, strong magnetic field, and many intriguing moons—some of which might even contain water or support life.

Pioneer

The journey to Jupiter began with NASA's Pioneer 10 and 11 in the early 1970s—our first robotic scouts to brave the outer solar system. Though expected to operate briefly, Pioneer 10 transmitted data until 2003, more than thirty years after launch.

The spacecraft were equipped with a variety of instruments, including magnetometers, particle detectors, imaging systems, and atmospheric analyzers. They incorporated innovative engineering features such as precision Doppler tracking, which helped with navigation and gravitational physics research. Pioneer 10 made history as the first spacecraft powered entirely by Radioisotope Thermoelectric Generators (RTGs), allowing it to travel beyond the reach of solar power.

Pioneer 10 was the first to cross the asteroid belt and actually observe Jupiter up close. Later, Pioneer 11 became the first spacecraft to encounter Saturn in 1979. Both missions paved the way for exploring interstellar space, showing that long-term deep space travel is possible.

Galileo

If Pioneer opened the door, Galileo stepped right through it. Launched in 1989, it became the first probe to enter Jupiter's orbit in 1995, sticking around for years of close observation. With it came the first atmospheric descent probe ever sent to another planet's clouds—measuring temperature, pressure, and chemicals inside Jupiter's swirling fog.

Galileo unveiled Jupiter's magnetosphere in stunning detail and explored how its fierce radiation interacts with the planet's many moons. It famously discovered volcanoes erupting on Io—the most volcanic body anywhere—and found tantalizing evidence of a salty ocean below Europa's icy shell, kept warm by tides. Europa instantly became a top target in the hunt for life beyond Earth.

Though expected to last two years, Galileo's mission stretched deep into the 2000s, forever transforming our picture of Jupiter as an active, everchanging world.

Juno

In 2011, NASA launched Juno, a solar-powered marvel built to peer beneath Jupiter's cloud tops. It reached Jupiter in 2016 with a unique polar orbit. Its instruments include microwave radiometers, magnetometers, particle detectors, and imaging systems for studying Jupiter's auroras. A protective titanium vault shields its electronics from Jupiter's punishing radiation.

Originally planned for about a year with thirty-seven orbits, Juno's mission has now been extended several times. It revealed that Jupiter's magnetic field is even more powerful, complex, and uneven than scientists had imagined. Beneath the clouds, hidden storms and jet streams swirl hundreds of kilometers below the surface, and Juno's data suggest that Jupiter's core is larger yet less dense than previously thought. These findings are challenging and expanding our understanding of how giant planets form and change over time. Plus, bonus flybys of moons like Ganymede and Io are giving us even more fascinating insights into the Jovian system.

Europa Clipper

Next comes Europa Clipper, NASA's ambitious mission planned for launch in the 2030s. Its goal is to investigate Jupiter's moon Europa for conditions suitable for life. It will study Europa's thick ice shell and the subsurface ocean believed to exist beneath it, examining geology, ice dynamics, and surface composition. The mission will also analyze water vapor plumes erupting from Europa's surface.

The spacecraft will carry advanced instruments including cameras, spectrometers, ice-penetrating radar, and magnetometers. Built to withstand Jupiter's harsh radiation environment, Europa Clipper uses Gallium Nitride (GaN) power electronics that operate efficiently at high voltages, temperatures, and switching speeds.

By conducting multiple flybys of Europa, the mission aims to determine whether this ocean world could harbor conditions for life, potentially redefining our understanding of where life might exist beyond Earth.

What We've Learned

Exploring Jupiter has opened a window into wild new territory, revealing weather and storms on a scale unmatched elsewhere in the solar system. Jupiter's immense gravity and relentless tidal forces spark volcanoes on Io and may keep hidden oceans warm and flowing beneath Europa's icy shell. Each mission–Pioneer, Galileo, Juno, and the upcoming Europa Clipper–rewrites our understanding of this swirling titan, pushing the limits of where life might exist and illuminating the wonders of our system's greatest giant.

But the story of Jupiter is far from over. Every robotic pioneer marks another chapter in our quest to understand the king of planets and its many moons. Armed with sharper eyes, tougher electronics, and bigger questions, the next generation of spacecraft will dive deeper than ever into Jupiter's swirling clouds, roaring storms, and mysterious satellites. What secrets lie beneath its cloud tops, and what lessons will Jupiter teach us about how planets–and maybe life–originate in other star systems?

As we unravel Jupiter's mysteries, we glimpse not only the vast possibilities of giant planets across the galaxy but also new insights into our own origin story–one waiting to be uncovered in the heart of the solar system's greatest wonder.

Chapter 45

Lord of the Rings

Saturn and Its Icy Rings

Saturn is the showstopper of the solar system, its dazzling rings making it look like an elegant plate-spinner balancing hoops of ice and stone. The sixth planet from the Sun and second only to Jupiter in size, Saturn is a true giant yet almost ethereal. Saturn is made mostly of hydrogen and helium, with barely any solid surface beneath those whipped-cream clouds. It's so light that—if you could find a bathtub big enough—Saturn would float.

With a volume that could hold over 760 Earths, Saturn's pale yellow color comes from ammonia ice in its upper atmosphere. Bands of clouds circle the planet in pastel stripes, driven by winds that howl at more than 1,000 miles per hour—faster than any storm on Earth. Over 270 moons—and hundreds more moonlets —orbit in Saturn's gravitational embrace, led by Titan, a mysterious world shrouded in smoggy haze and home to rivers and lakes of liquid methane—an Earth-like landscape, but with chemistry from an alien cookbook.

Voyager: The First Glimpse

In 1977, NASA launched the twin Voyager spacecraft—time capsules and trailblazers on a journey to outer planets and beyond.

Originally intended to study only Jupiter and Saturn, the Voyager missions were given an extraordinary opportunity when a rare planetary alignment allowed Voyager 2 to continue to Uranus and Neptune, greatly expanding our understanding of the outer solar system.

Both spacecraft carried advanced scientific instruments, including CCD cameras that revolutionized space photography with clearer, detailed images. They also featured spectrometers, magnetometers, and particle detectors to study planetary atmospheres and space environments.

Key innovations included their powerful communication systems with high-gain antennas capable of transmitting data across billions of miles via the Deep Space Network, and 3-axis stabilization that maintained precise orientation for observations.

Yet perhaps the most remarkable object onboard was the golden record, a twelve-inch copper LP plated in gold and attached to the side of each Voyager. Curated by a team led by the renowned American astronomer, planetary scientist, and science communicator Carl Sagan (1934-1996), it serves as both a time capsule and a universal greeting card, introducing our world to any curious extraterrestrials who might one day intercept the spacecraft. The record contains the sounds of Earth—a baby's cry, whale song, greetings in fifty-five languages, music from Bach to Javanese gamelan, and 116 images depicting scientific knowledge, nature, and daily human life. Engraved instructions explain how to play it, just in case the finders are deciphering our message across the gulf of both time and possibility.

Sagan famously called the golden record "a bottle into the cosmic ocean," a hopeful message that says as much about us as it does to any listeners. He wrote, "The spacecraft will be encountered and the record played only if there are advanced space-faring civilizations in interstellar space, but the launching of this 'bottle' into the cosmic 'ocean' says something very hopeful about life on this planet."

And, as Sagan reflected: "This is a present from a small, distant world, a token of our sounds, our science, our images, our music, our thoughts, and our feelings. We are attempting to survive our time so we may live into yours."

Voyager now carries this golden bottle into the stars, its music, sounds, and greetings sailing far beyond the Sun's reach as an enduring symbol of our desire to be known, and to know others, across the universe.

Though initially expected to operate for just five years, both Voyagers continue sending valuable data today. Voyager 1 entered interstellar space in 2012, crossing the heliopause where the Sun's influence ends, with Voyager 2 following in 2018—marking humanity's first exploration beyond our solar system.

The missions revealed active volcanoes on Jupiter's moon Io, complex structures in Saturn's rings, and provided insights into the atmospheres of Uranus and Neptune. By November 2026, Voyager 1 will be one full light-day (25.9 billion kilometers) from Earth.

Pretty cool to think that this little spacecraft, which was built with '70s tech and has less computing power than a modern smartwatch, is still out there, rocketing into interstellar space and carrying that golden record mixtape for any curious aliens it might meet. Not bad for something older than the first Star Wars movie!

Technologies developed for Voyager have found important Earth applications: image processing techniques now improve medical imaging in CT scans and MRIs. Miniaturization advances influenced modern electronics. Fault protection software pioneered concepts used in critical systems from nuclear plants to transportation networks.

Voyager's communications technology improved telecommunications systems, while its RTG technology has been adapted for remote power generation. Data analysis techniques developed for Voyager created new approaches to big data management in modern computing.

Cassini-Huygens: The Deep Dive

For more than a decade, Cassini orbited Saturn, weaving through its rings and moons, transforming our understanding of the planet's beauty and complexity. The Cassini-Huygens mission truly unlocked Saturn's mysteries. Launched in 1997, this NASA-ESA-Italian Space Agency collaboration spent thirteen years orbiting Saturn, completing 294 orbits and 162 moon flybys with advanced scientific instruments.

Cassini's radar pierced Titan's orange fog, mapping methane lakes and rivers on its frozen plains. In 2005, the Huygens probe made a parachute drop to Titan's surface—the first and only landing in the outer solar system to date—snapping panoramic images of a world as haunting as it is alien.

Cassini saw water vapor jets erupting from Enceladus, revealing a subsurface ocean and making it a prime candidate for extraterrestrial life. Its flybys uncovered new moons, unraveled the puzzle of Saturn's rings (a swirling mix of dust, pebbles, and icy boulders), and witnessed a bizarre six-sided storm at the planet's north pole—a weather pattern unknown anywhere else.

After dozens of daring passes between the planet and its main rings, and four mission extensions, Cassini's final act was a deliberate plunge into Saturn's atmosphere in 2017, ensuring no hitchhiking microbes could contaminate the planet's potentially habitable moons. Cassini's mission was a testament to engineering, teamwork, and a healthy dose of audacity.

What We've Learned

Thanks to these robotic explorers, Saturn is no longer just the "lord of the rings." It's a vibrant, ever-changing, and majestic planet—a grand, golden clockwork of winds, storms, and orbiting moons. Saturn's rings, which we once thought were solid, are a shimmering flow of frozen debris. Both Titan and Enceladus have captured our imagination as two of the most exciting places in the solar system to look for life—one with methane lakes under a nitrogen sky, and the other with watery plumes reaching out into space.

And the adventure with Saturn continues. Every new mission builds on the last, opening up more ways to peer beneath the haze, dive into the rings, and explore those mysterious moons. Saturn is more than just a distant world; it's a gateway to understanding planetary formation, the potential for habitability, and the endless creative spirit of the cosmos. Beyond those stunning rings, the upcoming missions promise even more amazing surprises and wonderful discoveries.

Chapter 46

Ice Giants

Uranus and Neptune Are Weird

Far out in the cold, blue twilight of the solar system, Uranus and Neptune—our "ice giants"—push the boundary between familiar worlds and the vast cold beyond.

If Jupiter is a spinning titan and Saturn rules the rings, these distant planets are the deep-sea mysteries: faint, icy, and challenging even for our bravest machines to explore. Their remoteness means missions require years of travel just to get a quick look—and so far, only one spacecraft has made the journey. Nearly everything we know about Uranus and Neptune comes from Voyager 2, which performed the only close flybys of these worlds in the 1980s. Before that, both appeared as ghostly blue dots in blurry telescopic images. But as Voyager 2 flew past Uranus in 1986 and Neptune in 1989, those ice giants came into focus as vibrant, dynamic planets—full of storms, moons, and mysteries never seen from afar.

Rolling Sideways

One of Voyager 2's most interesting findings at Uranus is its bizarre tilt. While Earth—like most other planets—spins like a top, Uranus rolls like a bowling ball around the Sun on its side, with an axis tilted about ninety-eight degrees.

For a quarter of its enormously long eighty-four-year journey around the Sun, one whole pole bakes in sunlight while the other freezes in night, creating bizarre, drawn-out seasons unlike anywhere else. Its pale blue-green glow comes from methane absorbing red light, its rings arch closer in than Saturn's, and Voyager spotted more moons than astronomers expected.

Uranus's magnetic field added another surprise: it's off-center, tilted, and skewed in ways unseen on any other planet—another sign of its deeply strange interior.

Wind Breaker

Neptune, Voyager's final planetary port of call, proved even more spectacular. The spacecraft measured the fastest winds in the solar system—over 1,200 miles per hour—racing across a world so distant that it receives only a faint trickle of sunlight. Neptune displayed a massive storm called the Great Dark Spot, a temperamental cousin to Jupiter's long-lived red one.

Its rings were patchy and clumped, more like scattered arcs than the smooth halos of Saturn. Voyager 2 also discovered new moons, including the irregular Proteus and the far more intriguing Triton. Triton orbits Neptune in a retrograde, backward direction and shoots plumes of nitrogen from its icy surface—strong evidence it was captured rather than born beside Neptune.

Neptune radiates more internal heat than Uranus despite receiving less sunlight, one of several deep mysteries still awaiting explanation.

Known Unknowns

The Voyager 2 mission gave us incredible snapshots, but only fleeting glimpses. Unlike Jupiter and Saturn, neither Uranus nor Neptune has had a dedicated orbiter or probe diving into its clouds. Most of their secrets—deep, swirling interiors, complex weather, and hidden chemistry—remain locked away, waiting for the next generation of explorers.

Uranus and Neptune are often called "ice giants," a type of planet that surprisingly appears quite common around other stars. Learning more about these cold, mysterious worlds would not only deepen our understanding of our own solar system but also help us decode the thousands of similar exoplanets discovered across the galaxy.

Every new glimpse offers exciting opportunities to learn how planets form, how atmospheres behave, and where habitable environments might be hiding among the stars.

All of this has sparked a strong push for dedicated missions to explore the ice giants. Uranus and Neptune are among the most intriguing planets in our solar system—cold, distant, and unlike anything we've studied before. Seeing them up close could really expand our knowledge of the solar system and shed light on how many ice giants orbiting other stars might form, evolve, and maybe even flourish in the universe's coldest regions.

Beyond the ice giants lies an even stranger frontier—the Kuiper Belt—yet the keys to understanding those frozen worlds may begin with Uranus and Neptune themselves.

Chapter 47

Dwarf Planets

Pluto and Beyond

Beyond the furthest reaches of the Sun, past the orbits of the giant planets, lies the Kuiper Belt—a vast, icy junkyard of ancient remnants from the birth of our solar system. It's in this frozen darkness that NASA's New Horizons spacecraft made history, becoming the first to explore Pluto and continue its journey onward. Launched in 2006, New Horizons traveled over three billion miles through nearly a decade of silence before reaching Pluto on July 14, 2015. Until then, Pluto was known as the enigmatic "ninth planet," a mysterious, dim dot scientists believed to be a sleepy, frozen relic. What actually unfolded was a planetary love letter—stunning high-definition images revealing a lively, vibrant world. Close-up photos showed a landscape far more dynamic than anyone had guessed. Bright nitrogen plains, including the famous heart-shaped Tombaugh Region, showcased mountains as tall as the Rockies, carved from solid water ice. Glaciers moved and shifted across the surface, suggesting geological forces are still at work. Even more exciting, plumes from ice volcanoes hint that Pluto, far from the Sun, is anything but quiet. New Horizons also showed us Pluto's delicate atmosphere—a thin halo of nitrogen rising hundreds of kilometers above the surface, with a blue haze that scatters sunlight in unexpected ways.

After reshaping our view of Pluto, New Horizons pressed onward into the Kuiper Belt, focusing its cameras on distant worlds that had only ever appeared as tiny pinpricks of light. Its first target was Arrokoth, a small, snowman-shaped object about 35 kilometers long, or 22 miles from end to end. Instead of being a single solid lump, Arrokoth is made up of two rounded "snowballs" of frozen material that seem to have gently come together and stuck, forming a narrow neck where they meet—quite like stacking two snowballs to create a snowman.

Up close, those "snowballs" aren't soft snow, but an ultra-cold mixture of water ice, other frozen chemicals, and a little rock, all tinged with a reddish-brown hue from complex organic materials. At temperatures below minus 200 degrees Celsius, this mixture is as hard as stone, helping the lobes maintain their smooth, lightly cratered surfaces for billions of years. This pristine, fused shape tells us that the two pieces drifted together at a gentle walking pace rather than colliding in a violent crash. It's a wonderful piece of evidence showing that some of the solar system's earliest building blocks grew through peaceful mergers of small, icy chunks in the dark outskirts of the Sun's disk.

Powered by a durable radioisotope thermoelectric generator, New Horizons keeps exploring the vast cosmos and sending back precious data. As it ventures farther into the outskirts of the Sun's realm —toward the distant boundary where the heliosphere thins into interstellar space—the spacecraft continues to sample, scan, and surprise us, its golden data stream revealing the secrets of the outer frontier.

The Kuiper Belt is a vast, doughnut-shaped region, stretching about thirty to fifty times farther from the Sun than Earth. It's home to thousands of icy remnants—proto-planets, leftovers, and comet seeds—that have remained frozen since the early days of our solar system. Many of these small icy objects are mainly composed of frozen methane, ammonia, and water—leftover materials from the solar system's formation, stored far away from the Sun's warmth.

The Kuiper Belt and its nearby scattered disc are also recognized as the birthplace of many short-period comets—icy visitors that occasionally travel into the inner solar system, shining brighter as they heat up near the Sun. Each tiny world, from Haumea and Makemake to Eris (whose discovery played a role in Pluto's reclassification), adds its own unique story of our solar system's history.

Losing Status

In 2006, astronomers made a big change to how we define planets. To be a planet, a world needs to orbit the Sun, be big enough for its gravity to make it mostly round, and clear the area around its orbit of other stuff. Pluto, which is round and orbits the Sun, shares its space with a bunch of other icy worlds. So, Pluto was reclassified as a "dwarf planet," like Eris, Haumea, and Makemake. But Pluto is still a cool and active world with lots of interesting landscapes and complex geology, not just a small, frozen rock.

But names have not diminished Pluto's wonders. Thanks to New Horizons, we see not a pale, exiled rock but a dynamic world, sculpted by geological forces and shaped by chance collisions and ancient leftovers long forgotten. Arrokoth and Pluto offer glimpses back to a dawn when the building blocks of planets came together—piece by icy piece—when the Sun was still young.

Now, New Horizons speeds away into the unexplored dark, a testament to what is possible when curiosity drives a spacecraft on an adventurous journey that never stops. It's still exploring within the Sun's enormous heliosphere, but each new image and every transmission from that distant, sunlit edge remind us that the solar system's story is far from over. In fact, some of the most exciting surprises await in the coldest, quietest regions, uncovered by one of the most successful and inspiring missions in space science history.

Signals

Out where the solar system frays,
The Kuiper Belt holds ancient days–
A frozen vault of cosmic debris,
Where New Horizons roams wild and free.

In 2006, this brave explorer flew,
Three billion miles through starry blue.
Nine years it sailed through space's night,
To gift us Pluto in brilliant light.

We thought we'd find a lifeless sphere,
A frozen rock–so cold and austere,
Instead? A world that's strangely alive–
With mountains tall where glaciers thrive.

Nitrogen plains stretch wide and far,
A heart-shaped region (quite bizarre!),
Ice volcanoes hint and tease,
At secrets hidden in the freeze.

Believing the mission was nearly done,
In 2019, New Horizons stunned everyone
A snowman-shaped cosmic stone,
The farthest world we've ever known.

Arrokoth showed us something grand–
How planets formed from dust and sand.
And though Pluto lost its planetary crown
(The IAU voted it down),

It couldn't clear its neighborhood, you see,
So "dwarf planet" it must be.
Yet size can't dim its strange appeal–
Its complex beauty is quite real.

Now onward still the spacecraft flies,
Through distant dark and silent skies
Sending signals from the dark unknown,
One of the farthest messengers we've ever flown.

Chapter 48

Cosmic Snowballs

Our Cosmic Time Capsules

Comets are some of the most captivating storytellers in our solar system—ancient travelers that streak across the sky with glowing tails, inspiring wonder for centuries. What makes them so unique compared to asteroids? Where do these cosmic snowballs come from, and what secrets might they hold about our own origins?

Fred Whipple, the same chap who designed the Whipple Shield, suggested in 1950 that comet cores are icy time capsules: a frozen blend of gas, ice, and dust formed far from the Sun in the earliest and coldest regions of the solar system—what he called a "dirty snowball." This set them apart from asteroids, which are mostly composed entirely of rock or metal. When these icy travelers approach the Sun, their ices vaporize and carry dust along, creating those stunning comas—the fuzzy cloud surrounding the nucleus—and tails that can stretch across millions of kilometers.

The "Dirty Snowball Model" explains why comets develop tails and become more active as they near the Sun, while remaining dark and quiet when they're far away in deep space—and it continues to be a key part of comet science today.

Comet Giacobini-Zinner

The modern era of comet exploration began in 1985, when NASA's International Cometary Explorer (ICE) performed history's first dive through a comet's tail—passing through the streaming plasma trail of Comet Giacobini-Zinner. ICE's instruments measured magnetic fields and tracked charged particles, revealing a massive, intricate tail stretching over 25,000 kilometers. This close encounter opened a brand new window for us to understand how comets interact with the Sun's hot breath—the solar wind.

Halley's Comet

The next act featured the most famous comet of all. When Halley's Comet returned in 1986, it became the focus of an unprecedented international fleet: Europe's Giotto, the Soviet Vega 1 and 2, and Japan's Sakigake and Suisei.

Giotto was the first to photograph Halley's dark, potato-shaped nucleus, revealing a surface so black it reflected only four percent of sunlight—a dusty, irregular surprise. The Soviet Vega spacecraft tested balloons in Venus's clouds before streaking past Halley, sampling its dust and gas. Japan's craft monitored its hydrogen shroud and solar wind interactions.

Together, these missions showed that comets are not static ice balls, but dynamic, active bodies whose tails, jets, and eruptions record the drama of their solar approach.

Rosetta and Philae

The European Space Agency's Rosetta probe, launched in 2004, advanced comet science into a new era. After a decade of pursuit, Rosetta orbited Comet 67P/Churyumov-Gerasimenko and deployed the determined Philae lander for the first-ever surface touchdown—though it was a bouncy, shadowed landing. Its suite of instruments found complex organic molecules and striking surface features: cliffs, pits, plains, and jets of gas and dust that seem to make comets "breathe" as they approach the Sun. Rosetta and Philae demonstrated that comet chemistry is more complex than we had ever thought, with implications for the origins of water and life on Earth.

Deep Impact

NASA's Deep Impact in 2005 added pyrotechnics to comet science. It launched a copper impactor to smash into Comet Tempel 1, blasting a crater and revealing the structure and chemistry beneath the surface. Spectrometers riding with the flyby craft measured the ejected material, confirming that comets are layered with porous, dusty crusts and tightly packed ices—an enduring record of their slow beginnings billions of years ago

What We've Learned

Thanks to these daring missions, scientists now see comets as cosmic laboratories—dynamic, even unpredictable worlds with striking surfaces, changing atmospheres, and "weather" shaped by sunlight. Their stunning tails and eruptions not only amaze the eye: they record solar winds, outgassing, and even solar storms.

Counting comet-related technological breakthroughs, missions like Giotto and Rosetta have contributed technologies used in advanced imaging and instrumentation, while the autonomous navigation technologies developed for deep-space missions are being adapted for robotics here on Earth. Understanding how comets break apart also helps us improve planetary defense against potential asteroid or comet impacts.

Perhaps the most exciting part is that comet dust contains organic molecules—life's fundamental chemicals—which supports the idea that early comets might have delivered some of Earth's vital building blocks. Every new mission, from Rosetta to New Horizons and beyond, broadens our understanding of the history of our solar system. Comets are no longer just "dirty snowballs"; they are active storytellers, guardians of ancient history, and maybe even messengers from the very beginnings of life.

Chapter 49

Dust Collectors

Capturing Comet Dust

Comet Wild 2 may seem like a pretty dirty snowball from afar, but when you get closer, it's full of surprises. Beneath its icy exterior, you'll find tiny grains of rock, crystals, and minerals like olivine and pyroxene– materials that normally form at very high temperatures near the young Sun (or even in other stars) before being transported outward. There's also a sprinkle of metallic iron-nickel compounds and even the building blocks of organic molecules. Each of these particles is like a little souvenir from different parts of the early solar system. Instead of just a snowball, Wild 2 acts like a wandering time capsule of our solar system's exciting beginnings.

In 1999, NASA's Stardust mission set out on an incredible journey– trying something no spacecraft had done before: flying through a comet's coma, collecting dust grains zipping by at hypersonic speeds, and bringing them safely back to Earth. No one had attempted to catch particles moving faster than a rifle bullet and keep them intact for study.

Wild 2 is a small, rugged icy body orbiting the Sun, covered in a drifting veil of vapor and dust. The tricky part was collecting particles racing at more than 14,000 miles per hour–fast enough to vaporize almost anything on contact. To solve this, Stardust used a special collector filled with aerogel, that's so porous particles can tunnel into it, leaving delicate tracks behind before finally coming to rest. The comet dust stayed intact like insects preserved in amber.

For nearly seven years and over three billion miles, Stardust journeyed through the dark reaches of space toward Wild 2, flew through its expanding halo, and then made the return trip home. In January 2006, a shimmering sample capsule parachuted into the Utah desert, carrying thousands of pristine comet grains from beyond the Moon's orbit—the very first time humans had collected material from a comet.

Discoveries and Insights

Stardust celebrated several firsts: it was the first to bring back cometary material, the first samples ever collected from deep space, and the first to capture interstellar dust as it drifted through our solar neighborhood. Along the way, it also enjoyed a bonus flyby of asteroid 5535 Annefrank in 2002. Instead of retiring, the spacecraft was given a bonus mission. In 2011, NASA repurposed it to revisit Comet Tempel 1 and inspect the aftermath of a cosmic hit-and-run: the massive crater blasted into the comet years earlier by the Deep Impact mission.

The samples returned by Stardust really surprised us. Hidden within the aerogel were minerals that formed at incredibly high temperatures near the young Sun—now found at the chilly edges of our solar system. This discovery showed us that the early solar system was much more lively and energetic than we had imagined, with materials being mixed violently across vast distances. The grains from Stardust also revealed an intriguing mix of organic compounds and a rich variety of chemicals, supporting the idea that comets could have delivered some of the very ingredients essential for life on Earth.

The mission relied on innovative technology that seemed straight out of science fiction: aerogel collectors, a dust counter that recorded every tiny impact, miniature chemistry labs, and navigation cameras that steered the craft through clouds of cosmic debris. The sample-return capsule was a marvel of engineering, designed to survive a fiery descent through Earth's atmosphere and landing safely delivering its priceless cargo.

Legacy and Impact

Led by Principal Investigator Dr. Donald Brownlee from the University of Washington, along with teams from NASA's Ames Research Center, Johnson Space Center, and institutions around the world, Stardust revolutionized the landscape of sample-return science. Its incredible achievement inspired a new wave of explorers eager to bring back pieces of our origins—missions like OSIRIS-REx, Hayabusa2, and future comet samplers all draw inspiration from Stardust's pioneering tools and design ideas.

Techniques using aerogel, dust collectors, and return capsules, first proven by Stardust, now serve as models for studying planetary materials across the solar system. With every grain of dust, Stardust helped tell a bigger story: how planets form, how water and organics travel between worlds, and how even the oldest materials can shed light on the solar system's forgotten past. Its legacy continues not only through the samples scientists analyze but also through the curiosity it reignited and the technologies it handed to humanity—powerful tools for asking questions that no one had dared to before Stardust brought a piece of a comet home.

Chapter 50

Star Watchers

Looking Out

Our journey into space started very close to home. At first, we looked down—gazing at clouds, watching the weather, and studying Earth from above. But soon, our curiosity and outlook grew outward. The same rockets that launched weather satellites began carrying telescopes and scientific tools deeper into space.

These new instruments broadened our view beyond Earth. They explored stars, galaxies, and distant planets. Some measured the faint afterglow of the Big Bang, while others captured stunning images of huge storms swirling across Jupiter and Saturn.

These new observatories allowed us to glimpse the universe in incredible new ways—detecting invisible forms of light, from radio waves to gamma rays. Together, they revealed a universe that is expansive, ancient, and wonderfully full of mysteries.

And it all began with an antenna listening to the quiet hiss of the stars. This exploration isn't just about satisfying curiosity. These missions have deeply changed the way we see our place in the universe. They've shown us how stars are born, live, and die, how galaxies form and grow, and how incredibly special our own planet truly is—in the vastness of the cosmos, Earth is just a tiny speck.

The amazing technologies created for space exploration have also brought wonderful benefits back home. Tools designed to study distant stars now help doctors detect illnesses, improve communications, and make imaging clearer. Computer programs that refine images from space telescopes assist in better medical scans, while advanced cooling systems and detectors improve how scientists store sensitive materials, including medicines and vaccines.

In the pages ahead, we'll explore two amazing ways we look into the universe: with radio telescopes that listen to the cosmos, and space telescopes that give us crystal-clear views from above Earth's atmosphere. These incredible instruments have transformed our understanding of space and continue to reveal breathtaking discoveries. They have reshaped not just what we know, but also the big questions we ask about our place in the universe.

Jansky's Merry-Go-Round

Throughout history, our understanding of the universe mainly depended on visible light—the colors and brightness of stars and planets we can see with our eyes or telescopes. Early astronomers like Galileo used glass lenses and mirrors to magnify this faint glow, creating telescopes to explore the sky. But looking at the universe only through visible light is like listening to a symphony by focusing on just one instrument—so much of the music goes unheard.

That all changed in the early 1930s, thanks to Karl Jansky (1905-1950) , a young engineer at Bell Labs who was investigating radio static that disrupted transatlantic phone calls. He built a strange, rotating one-hundred-foot-long antenna on a platform of old Ford wheels. Local engineers joked it was "Jansky's merry-go-round." The rotating antenna could aim in any direction, helping Jansky to track the source of the radio noise. After months of attentive listening, he identified three kinds of static. Two of them came from thunderstorms—some nearby, some far away. But the third was really exciting: a steady hiss that cycled every twenty-three hours and fifty-six minutes, which is exactly how long Earth takes to rotate once relative to distant stars. In other words, this hiss came from beyond our solar system—specifically, from the Milky Way galaxy, humming away in radio waves, unveiling a universe we couldn't see before.

Jansky's discovery was groundbreaking, and it was Grote Reber—an enthusiastic, self-taught radio astronomer—whose vision and curiosity pushed the field forward. In 1937, Reber built the first dedicated radio telescope: a 31-foot dish in his backyard in Wheaton, Illinois.

His pioneering efforts mapped out the invisible radio sky, opening the door to a whole new exciting era in astronomy.

The World's Great Radio Observatories

Today's radio telescopes are genuinely incredible feats of engineering. They feature large metal dishes, sometimes hundreds of feet in diameter, to collect and concentrate faint radio waves from the depths of space. These signals are incredibly faint—by the time they reach Earth, their energy is even less than that of a single falling snowflake!

Among these giants standing watch beneath our sky, the Robert C. Byrd Green Bank Telescope (GBT) in West Virginia stands out as the largest fully steerable radio telescope in the world. It features a hundred-meter-wide dish mounted on a complex wheel-and-track system. This mighty structure, weighing around 17 million pounds and reaching heights equivalent to a forty-story building, can point to nearly any part of the sky, helping us map the universe in exquisite detail. Situated in one of Earth's quietest radio zones, the GBT catches faint cosmic signals—from pulsars, star nurseries, and the swirling centers of galaxies to fleeting bursts that could come from across the universe.

Global efforts in radio astronomy have advanced with telescopes like the Very Large Array (VLA) in New Mexico—a vast "Y" of twenty-seven movable dishes stretching about twenty-two miles—and South Africa's MeerKAT, which enhances sensitivity even further. These arrays combine their signals to produce detailed radio "images," uncovering hidden structures, energetic jets, and intricate maps of the universe's magnetic fields. They have charted star formation in cold gas clouds, examined ancient radiation from the early universe, and tracked fast radio bursts (FRBs).

Radio astronomy is on the verge of its most exciting breakthrough yet. The Square Kilometre Array (SKA), being built in South Africa and Australia with international cooperation, will use thousands of antennas spread over hundreds of kilometers. Once finished, the SKA will have a total collecting area of about one square kilometer, making it one of the largest and most sensitive radio telescope arrays ever created. Its sheer scale and advanced design will allow scientists to scan the sky much faster and detect much fainter signals, helping us explore the universe's "dark ages," discover where the first stars and galaxies came from, map cosmic magnetism, and even search for signs of life beyond Earth.

Thanks to pioneers like Jansky, Reber, and countless innovators, today's radio telescopes enable us to "hear" more of the universe's symphony–revealing a richer and more detailed story about the cosmos and our place in it. The universe communicates through much more than visible light, and at last, we are learning to listen.

What Radio Telescopes Have Taught Us

Radio astronomy has really revolutionized our understanding of the universe with incredible discoveries like blinking pulsars, shining quasars, and, most famously, the unexpected detection in 1964 by Arno Penzias and Robert Wilson of the Cosmic Microwave Background (CMB) radiation. This faint, steady radio "hiss" turned out to be the ghostly afterglow of the Big Bang–a remnant heat signal from the universe's birth, still echoing across space and time. By using radio telescopes, astronomers have mapped unseen networks connecting galaxies, teaching us to see the cosmos not just as a silent void, but as a vibrant symphony across many frequencies.

Sometimes, embarking on the journey of discovering new knowledge can come with a bit of human drama. Back in 1967, Dame Jocelyn Bell Burnell, then a young graduate student at Cambridge in 1967, was the first to notice the regular, rhythmic signals that we now call pulsars–spinning neutron stars that shine like cosmic lighthouses. With persistent effort through countless chart recordings, Bell Burnell identified a tiny, repeating signal known as "scruff," which exhibited extraordinary regularity. Her unwavering dedication led to one of the most exciting astronomical discoveries of the 20th century, opening up a whole new chapter in astrophysics and offering indirect proof of neutron stars and other extraordinary cosmic phenomena.

Although Bell Burnell played a crucial role, she was not named in the 1974 Nobel Prize in Physics, which was instead awarded to her supervisor Antony Hewish and Martin Ryle. This omission sparked ongoing conversations about fairness, recognition, and diversity in science. Always gracious, Bell Burnell has become a passionate advocate for women and underrepresented groups in astrophysics. She is celebrated as a pioneering scientist and role model whose perseverance and dedication continue to inspire many.

Today, new generations of radio telescopes, enhanced by advanced computing and artificial intelligence, are ready to analyze vast amounts of data in search of exciting new discoveries. Somewhere out there, we might still detect a signal that could completely change our understanding of the universe.

Hubble Era

For centuries, Earth's atmosphere has made it difficult to get clear views of stars, planets, and distant galaxies, even with the most advanced observatories. Climbing to higher, thinner, and more stable air—like on Mauna Kea in Hawaii, over 4,200 meters tall, or Mount Wilson in California, about 1,742 meters high—improved things but never completely eliminated the atmosphere's constant turbulence. This phenomenon, called "atmospheric seeing," makes it feel like looking up through rippling water, and it limits the detail and sharpness of observations. The space age brought a big breakthrough: launching telescopes above the atmosphere to see unblurred starlight from space.

Enter the Hubble Space Telescope (HST), a project that took more than twenty years to develop. Its idea started back in the 1940s when Lyman Spitzer suggested putting a telescope in orbit to avoid the distortion caused by Earth's atmosphere. The telescope was named after Edwin Hubble (1889-1953), a pioneering American astronomer whose work transformed how we understand the universe.

After serving in World War I, Edwin Hubble worked at California's Mount Wilson Observatory, where in the 1920s he made two groundbreaking discoveries. First, he showed that "spiral nebulae" are actually distant galaxies beyond the Milky Way, greatly expanding our view of the universe. Then, in 1929, he described what became known as Hubble's law: galaxies are moving away from us, and the farther they are, the faster they seem to recede, providing key evidence that the universe is expanding. These discoveries formed the foundation of the Big Bang theory and modern cosmology.

Hubble also created the "Hubble tuning fork" classification system, which astronomers still use—with some updates—to organize galaxies by their shape. By opening up new frontiers, Hubble's work shifted astronomy from just studying our local stellar neighborhood to exploring an infinite, ever-changing universe.

The HST was launched into orbit in 1990 aboard the Space Shuttle Discovery, entering an orbit roughly 540 kilometers above Earth. Free from atmospheric distortions, it circles the planet roughly every 97 minutes. Its power comes from solar panels, and it uses highly precise gyroscopes and control systems to stay focused on its targets. The telescope features a primary mirror that's 2.4 meters across, polished to such perfection that, if scaled up to the size of the continental United States, the biggest bumps would only be a few centimeters tall. This incredible precision helps Hubble capture light from objects that are billions of light-years away, allowing us to explore the universe in amazing detail.

Tying it All Together

Trouble started almost right away when a tiny flaw in the primary mirror's shape blurred Hubble's images, leading to headlines about a potential billion-dollar blunder.

In December 1993, the Space Shuttle Endeavour took on its first servicing mission (STS-61) to restore Hubble's vision. Astronauts Story Musgrave and Kathryn Thornton installed a special corrective optics package called COSTAR—like prescription glasses for Hubble's mirror—that fixed the blurring and brought back its clarity. In a display of NASA's problem-solving spirit, Musgrave used parachute cord to secure a critical thermal blanket that ensured HST was protected from the extreme temperatures of space.

From that point on, HST has transformed faint smudges into stunning, sharp images of galaxies racing apart, vibrant nebulae, and the birth, life, and death of stars that were once hidden from view. Its powerful scientific instruments are capable of capturing visible, ultraviolet, and parts of infrared light, allowing us to see everything from sweeping deep fields of distant galaxies to close-ups of newborn stars and planets forming in dusty disks.

Hubble's Stunning Accomplishments

Hubble's incredible contributions are woven into the fabric of modern astronomy. It has helped refine our understanding of the universe's age, shown how galaxies grow and collide, and traced the light from exploding stars to explore dark energy. Hubble has also discovered moons and weather patterns on distant planets and contributed to the study of exoplanet atmospheres. Beyond pure science, its advancements in optics, image processing, and vibration control have inspired technologies like high-resolution imaging systems and stable platforms for delicate instruments both on Earth and in space.

The breakthroughs in data handling and image processing have empowered powerful algorithms that benefit fields like medical imaging and climate modeling. Its position above the atmosphere allowed Hubble to capture breathtaking images—such as the iconic "Pillars of Creation" in the Eagle Nebula and deep field galaxy surveys—that really changed how we see the universe. Hubble's influence continues to inspire countless students, artists, and scientists, nurturing new generations of explorers and creative minds.

Today, Hubble is much more than an incredible scientific instrument—it's a shining symbol of what can be achieved through vision, perseverance, and teamwork. Its legacy goes beyond the breathtaking images that fill books, posters, and screens; it helps us better understand our place in the universe—and reminds us that gazing up at the sky is always a worthwhile and inspiring adventure.

James Webb Generation

The James Webb Space Telescope (JWST) stands as an exciting new flagship in our scientific community, symbolizing one of humanity's most amazing efforts to explore the cosmos and uncover its ancient stories. Named in honor of James E. Webb (1906-1992), NASA's second administrator who played a key role in the Apollo program and expanded NASA's scientific pursuits, JWST carries forward the incredible legacy of the Hubble by peering into the infrared universe.

The idea for JWST was born in the mid-1990s, when scientists such as John Mather and others envisioned a telescope that could look back in time to see the earliest galaxies formed after the Big Bang. Unlike Hubble, which orbits relatively close to Earth, JWST operates from the second Lagrange point (L2)—about 1.5 million kilometers away—where the combined gravity of Earth and the Sun creates a stable environment perfect for observation. This unique placement helps keep the telescope's instruments both cool and steady, allowing it to do its incredible work.

A Telescope to See Back in Time

Webb's mission is simple to state yet ambitious in scope: to explore where galaxies, stars, and planets come from; to look through cosmic dust to find the first light that brightened the universe; and to study distant worlds for signs of atmospheres and habitability. Its main feature is a gold-coated primary mirror, 6.5 meters across, made of eighteen hexagonal segments arranged in a honeycomb pattern that unfolds in space. Unlike Hubble's mirror, which is tuned for visible and ultraviolet light, Webb specializes in infrared, a key tool for seeing ancient light stretched by the universe's expansion and piercing through dust to reveal newborn stars.

This sensitivity requires Webb to stay extremely cold—only a few degrees above absolute zero for some instruments. A tennis-court-sized, five-layer sunshield blocks heat from the Sun, Earth, and Moon, keeping the observatory in a frigid environment so it can detect faint signals from the distant universe. Webb's MIRI instrument is cooled even further, to about 7 Kelvin, by a dedicated cryocooler so it can observe at longer infrared wavelengths.

Engineering Marvels

JWST's suite of scientific instruments brings together some of the boldest engineering ever launched into space:

Near Infrared Camera (NIRCam): Led by Marcia Rieke and her team, NIRCam serves as Webb's primary imager. It looks far back in time, capturing "first light" from early stars and galaxies and tracing how cosmic structure emerged from the universe's long "dark ages."

Near Infrared Spectrograph (NIRSpec): Designed with major contributions from Peter Jakobsen and the ESA, NIRSpec uses a programmable microshutter array–thousands of tiny individually controlled shutters–that allow astronomers to observe the spectra of up to about one hundred objects at once.

Mid-Infrared Instrument (MIRI): Developed through European–American collaboration including work led by Dr. Gillian Wright, MIRI operates at very low temperatures, just 7 Kelvin, and can image intricate structures of cosmic dust, star-forming regions, planetary disks, and extremely distant galaxies.

Fine Guidance Sensor / Near InfraRed Imager and Slitless Spectrograph (FGS/NIRISS): Provided by the Canadian Space Agency, FGS keeps Webb pointed with extraordinary accuracy, while NIRISS supports detailed studies of exoplanet atmospheres and other specialized observations.

Collectively, these incredible instruments enable Webb to peer deeper into the universe's history than any telescope before it. This allows us to discover new insights about how stars, planets, galaxies, and even life's potential conditions come into being.

Unlike Hubble, which astronauts could repair if needed, Webb is stationed far beyond our immediate reach, so it had to be designed to unfold and operate smoothly on its own, ensuring it can do its amazing work without any hands-on assistance. Webb's first images revealed galaxies from the universe's early epochs, mapped exoplanets and their atmospheres, and provided some of the sharpest infrared views of nebulae and star-forming regions ever obtained. In a short time, it has already reshaped several areas of astrophysics and continues to do so with each new data release.

From the Edge of Space to Everyday Life

JWST's innovations are also making a positive impact on technology here on Earth. Its segmented beryllium mirror, ultra-stable structures, precision pointing, cryogenic systems, and micro shutter arrays represent advances in materials, optics, and micro-electromechanical systems. These technologies and the associated testing methods are helping in fields like semiconductor fabrication, high-performance imaging, and cryogenic engineering, while also guiding the development of future scientific instruments and industrial systems. Plus, the large-scale data processing and algorithms created for Webb's data are driving broader progress in data science and artificial intelligence.

What Comes After Webb

The success of JWST paves the way for even more ambitious telescopes in the future. LUVOIR, or the Large UV/Optical/IR Surveyor, is a concept for a future space telescope with a mirror potentially up to about 15 meters across–several times larger than Webb's. It's designed to give us even more stunning views of exoplanets and distant galaxies. NASA's upcoming Nancy Grace Roman Space Telescope, scheduled to launch later in the 2020s, will have a field of view roughly 100 times larger than Hubble's. This will allow scientists to survey huge areas of the sky quickly, helping to map dark matter and dark energy while discovering thousands of exoplanets.

The Habitable Exoplanet Observatory (HabEx) is another concept that aims to directly image Earth-sized planets around other stars, using a large telescope paired with a starshade–a giant, flower-shaped shield that blocks the glare of a star so faint planets can be seen. Beyond these, visionary ideas include enormous observatories made from swarms of smaller spacecraft flying in formation, telescopes built on the Moon's far side and discover thousands of exoplanets along the way.

Dark Matter

Gaia is one of the most incredible star-mappers—a European Space Agency (ESA) observatory launched in 2013 to create the most detailed and accurate 3D map of the Milky Way ever attempted. The name "Gaia" originally stood for the Global Astrometric Interferometer for Astrophysics, an early idea for an interferometric instrument. It also echoes Gaia, the Earth goddess from Greek mythology—an apt symbol for a mission dedicated to exploring our galactic neighborhood.

From its parking spot near the Sun—Earth L2 Lagrange point, about 1.5 million kilometers away from Earth—Gaia set out to measure the positions, distances, and movements of roughly one billion stars, about one percent of all the stars in our galaxy. That's more than all previous star catalogs combined! Over more than ten years of scanning, Gaia has charted not just where stars are, but how they move, how fast they travel, and how their paths weave through the spiral arms of the Galaxy.

Gaia's measurements are so precise that it can detect tiny shifts in the sky as small as tens of microarcseconds—comparable to spotting a human hair from hundreds of kilometers away. With this incredible accuracy, Gaia has revealed the complex structure of the Milky Way, from spiral arms and star clusters to evidence of past mergers with dwarf galaxies. This helps astronomers reconstruct how our galaxy has grown and changed over billions of years.

Throughout its journey, Gaia has uncovered new worlds by detecting tiny stellar wobbles and timing changes. It has also tracked asteroids and comets to enhance planetary defense and cataloged distant quasars that act as fixed reference points for mapping the sky. The data collected has helped astronomers understand how dark matter—the invisible stuff that influences our galaxy's gravity—is spread across the Milky Way. Gaia's discoveries have become crucial for exploring everything from the history of stellar explosions to testing Einstein's General Theory of Relativity.

Gaia wrapped up its main science observations on January 10, 2025. After a few weeks of onboard tests, ESA powered down Gaia's systems on March 27, 2025. Then, using its thrusters one last time, the spacecraft was moved from its orbit at L2 into a safe, stable heliocentric "retirement" orbit, safely away from Earth. It was then passivated and officially decommissioned, so it won't interfere with future missions.

But even in retirement, Gaia's treasure chest of data—covering nearly two billion stars, exoplanets, asteroids, and more—will keep fueling discoveries for many years. No other mission has given us such a clear, detailed picture of our home galaxy or changed how we see our place in the universe as profoundly as Gaia. Its legacy will shine through every sky map and model of our galaxy for generations to come, helping us understand the story of the stars better and inspiring new questions along the way.

Chapter 51

Space Law

A Constitution for Space

When Sputnik was launched in 1957, humanity's first step into space, grabbing headlines and sparking a big question: who would be in charge of this new frontier? During a time of Cold War tensions, it became obvious that without proper rules, space could turn into a new battleground. But despite their rivalry, the United States and the Soviet Union both shared a common worry—an uncontrolled space could pose a danger to everyone.

Back in 1966, countries around the world came together to set some important rules for space exploration. The United States was mainly focused on protecting celestial bodies like the Moon, while the Soviets wanted to make sure that all space activities had proper regulations. After several months of negotiations—led by UN Secretary-General U Thant—everyone agreed on the 1967 Outer Space Treaty, officially called the "Treaty on Principles Governing the Activities of States in the Exploration and Use of Outer Space, including the Moon and Other Celestial Bodies." It was opened for signatures on January 27, 1967, in Washington, London, and Moscow, and officially became active on October 10, 1967. Today, it still serves as the key foundation for international space law.

This "constitution for space" set out a clear vision. Space exploration was to promote peace, scientific discovery, and the interests of all humankind. Its most important rule was non-appropriation: no nation could claim sovereignty over celestial bodies, preserving space as a shared global domain rather than a new theatre for colonization. To support that vision, the treaty emphasized the peaceful use of outer space and placed responsibility and liability for national activities firmly on states, whether carried out by governments or private companies.

From Treaties to Transactions

Four core principles from the Outer Space Treaty continue to shape everything from Mars missions to asteroid mining.

Non-appropriation means no country, company, or individual can claim ownership of the Moon, Mars, asteroids, or any part of outer space. Anything extracted is akin to gathering shells on a public beach: the shells may be yours, but the beach belongs to everyone.

Peaceful use bans weapons of mass destruction in orbit, on celestial bodies, and elsewhere in outer space, and forbids military bases, weapon tests, or military maneuvers on celestial bodies, but does allow military personnel and equipment to be used for scientific research or other peaceful activities, such as exploration.

Benefit to all requires that space activities serve the interests of all countries, regardless of wealth or technological sophistication.

Responsibility and liability make nations accountable for all national activities in outer space and internationally liable for damage caused by their space objects.

For decades, these treaty-level principles were sufficient for a world dominated by government-led scientific missions. As more countries and companies entered orbit and began pushing beyond it, gaps and ambiguities emerged. The 1972 Liability Convention clarified how fault would be assigned in accidents involving space objects, while the 1979 Moon Agreement attempted to address resource extraction and "common heritage" questions. That agreement, however, was never widely adopted by major spacefaring nations.

The tension between lofty principles and commercial ambition became sharply visible in cases such as that of Dennis Hope. In 1980, he filed paperwork claiming ownership of the Moon and founded the Lunar Embassy Corporation, selling "lunar deeds" for roughly twenty dollars per acre. Buyers received official-looking certificates for plots on the Moon, Mars, Venus, and even the moons of Jupiter. These deeds have no legal standing. International law is explicit that no person, company, or nation can own any part of a celestial body, and such claims are treated as novelties rather than property rights. Their popularity, however, reveals how powerfully the idea of owning a piece of space captures the human imagination.

Today, more than a hundred countries operate or participate in space programs, and private spaceflight has turned launches into routine events rather than rare state spectacles. Thousands of satellites now crowd Earth orbit. SpaceX alone has deployed thousands of Starlink satellites. Companies are planning to mine asteroids and the Moon, while scientists map lunar water deposits for future bases. Business plans speak openly of "space mining rights" and "orbital real estate," shifting space from a purely scientific frontier into a competitive commercial environment.

As commercial launches and resource-extraction concepts gathered momentum in the 2000s and 2010s, legal debates intensified. National laws in the United States, Luxembourg, Japan, and other countries began granting companies ownership of the resources they extract, while stopping short of asserting sovereignty over the territory itself. In 2020, the United States and partner nations introduced the Artemis Accords, a set of non-binding principles for peaceful lunar exploration and resource use among participants in NASA's Artemis program. More than two dozen countries have since signed, including Japan, Canada, and many EU states, while other major spacefaring nations have remained outside the framework. The underlying point remains simple. You cannot own land on the Moon or Mars, but several nations now say you can own what you bring back. Law is evolving in real time on genuinely new terrain.

Inviting the Private Frontier

In the 1980s, the United States Congress began constructing a commercial space framework that focused as much on encouragement as on restriction. The goal was to bring private operators into orbit and regulate them only to the extent required for safety and national interests, positioning law as an invitation rather than a barrier.

Roughly two decades later, as serious plans for flying paying passengers emerged, lawmakers added a second layer. Commercial human spaceflight was treated as both promising and fragile. Regulators were directed not only to permit the industry, but to actively promote its development while safeguarding public safety. Human spaceflight by private companies moved from the margins of the law to its center.

Congress also recognized that no one yet knew what "normal" would look like for this technology. Instead of locking in permanent standards too early, lawmakers designed a system in which safety rules would evolve alongside experience. The bargain was explicit. Regulation should not suffocate innovation, but it could not stand still as knowledge improved and risks became clearer. Safety and innovation were framed as forces to be balanced, not traded away.

To allow that balance to emerge, legislators created a learning period for commercial human spaceflight. Aviation-style certification was deliberately delayed while vehicles, operations, and experience matured. In those early years, the regulator's role was to observe, gather data, and intervene rather than impose a fully developed regulatory regime from the outset.

Taken together, these choices defined a relationship. The state asked private operators to take disciplined risks on its behalf, and regulators were expected to act as both safety guardians and enablers of ambition. Episodes such as a Virgin Galactic test flight briefly straying outside New Mexico airspace and being granted more room to operate, rather than less, make sense in this context. They are small, human-scale expressions of a system designed to keep people alive without extinguishing the urge to explore.

Pressure Points

Modern dilemmas such as extraction rights, base locations, and satellite liability highlight the urgency of updated frameworks. Competition for resources could drive multiple actors toward the same deposits of lunar water or rare minerals. The treaty's non-appropriation clause does not define "operational zones," opening the door to modern land-rush dynamics disguised as safety buffers.

Base placement raises similar concerns. While the treaty prohibits territorial claims, it says little about how exclusion zones around permanent facilities should be governed. Long-term bases will require local safety zones, but without agreed limits, those zones risk becoming de facto territorial claims—the very outcome the treaty sought to prevent.

The growth of commercial space activity also complicates liability and insurance. The 1972 Liability Convention predates mega-constellations, autonomous spacecraft, and continuous debris tracking. Space law cannot simply borrow from maritime or terrestrial precedents. Space has no global registry, no dedicated courts, and no natural boundaries to fall back on. Without coherent international solutions, the risks of conflict, confusion, and unmanaged competition will grow.

These challenges mirror broader struggles in global cooperation. If humanity cannot fairly manage space resources, it raises uncomfortable questions about our ability to govern other shared systems such as the climate, the oceans, or the Arctic. How space is handled today may shape international cooperation far beyond orbit.

When Space Gets Crowded

Space was once perceived as pristine, an endless black expanse that inspired awe. Today, Earth's orbital highways are crowded. Low Earth orbit, stretching roughly from 160 to 2,000 kilometers above the planet, has become a dense environment filled with fast-moving artificial objects of every size.

Since Sputnik, thousands of launches have left behind a growing debris field. More than 30,000 objects larger than about ten centimeters are actively tracked, while millions of smaller fragments move invisibly at roughly 28,000 kilometers per hour, each capable of damaging or destroying a spacecraft. In 1978, NASA researcher Donald J. Kessler warned that cascading collisions could render parts of orbit unusable for decades. This scenario, now known as the Kessler Syndrome, has shifted from theoretical concern to practical risk.

The 1972 Liability Convention makes launching states internationally liable for damage caused by their space objects, regardless of age or operational status. In practice, enforcement has been rare. The clearest test occurred in 1978, when the Soviet nuclear-powered satellite Cosmos 954 re-entered over Canada, scattering radioactive debris. After extended negotiations, the Soviet Union paid $3 million Canadian dollars toward cleanup costs.

Debris mitigation today relies largely on guidelines rather than binding law. The Inter-Agency Space Debris Coordination Committee recommends practices such as controlled re-entry or disposal into "graveyard orbits." Without legal force, however, these measures can be sidelined when budgets or schedules tighten, allowing congestion to increase with every launch.

Private companies now drive much of the new space race. SpaceX, Amazon, OneWeb, and others are deploying vast satellite constellations to provide global connectivity. Starlink alone envisions tens of thousands of satellites, each adding complexity to collision avoidance and traffic management. When satellites collide, questions of responsibility, jurisdiction, and insurance quickly become legally tangled. Regulation is struggling to keep pace with deployment.

Space Lawyers

As humanity moves from brief visits to permanent activity beyond Earth, legal systems face environments they were never designed to govern. Space law is shifting from abstract treaties toward practical rules for daily life and commerce in remote, high-risk settings.

Consider a future lunar scenario. International teams operate side by side. NASA crews mine resources. ESA scientists run experiments. Private companies construct habitats. A mining rover accidentally disables the power supply to a nearby research station. The result is not merely equipment loss but a crisis threatening lives, research, and infrastructure worth billions. On Earth, such incidents are governed by mature liability regimes and emergency protocols. In space, those systems remain incomplete.

The Outer Space Treaty was drafted for an era of government exploration and scientific cooperation. It prohibits sovereignty claims but leaves private activity, resource rights, and mixed public-private operations largely undefined. This ambiguity raises practical questions. How can extracted resources be claimed without violating non-appropriation? How is fault assigned when autonomous systems from different operators interfere with one another?

Emergencies off Earth also unfold under severe constraints. Rescue delays, limited redundancy, and communication lags demand new approaches to liability and insurance. Space law must become operational, capable of handling contracts, claims, and emergency response in real time rather than relying solely on high-level principles.

Responsibility, Not Ownership

As space becomes more accessible beyond Earth, space law is also evolving to keep up. A new wave of space lawyers—experts tackling challenges never faced on our planet—are playing a key role. Leaders like Frans von der Dunk and Joanne Gabrynowicz, along with research teams at Leiden, McGill, and the University of Mississippi, are helping shape this exciting new field.

Their focus is on practical solutions—think contracts for international bases, sharing resources fairly, and creating liability rules that work with existing treaties. In the future, some of the biggest legal debates might happen not in courts on Earth, but in lunar bases, space stations, or even virtual spaces across the solar system.

Most early cases are likely to be about everyday issues like equipment breakdowns, satellite crashes, power outages, and insurance claims. These incidents will set important precedents, helping us understand responsibility and accountability as we explore this new frontier.

Just as maritime law developed to govern the oceans, space law is coming into being to manage activities beyond our planet. It has to balance commercial interests with safety, encourage innovation while protecting the environment, and find ways to serve both national and global interests.

Now, space law stands at an important crossroads. It's not about who owns space anymore, but about how we share the responsibility for it.

The universe remains open to everyone. What we build and how we do it will depend on the values and standards we choose to uphold.

Wrap-Up

✦ ·✦· ✦

Your Space-Based Daily Life

We began this journey with a simple question: *what can space do for Earth?*

Now, as we reach the end, we see that space has given us something far more valuable than technological advances. It has given us a new way of seeing ourselves and our planet.

From the first rockets that pierced our atmosphere to the ecosystems we're designing for Mars, each step into space has taught us something profound about life on Earth. The challenges of surviving in space—managing resources, recycling everything, working together in tight quarters—are exactly the skills we need to build a better world here at home.

Look around you. The phone in your pocket uses technology developed for satellites. Atomic-clock timing keeps finance and power grids in sync and makes your maps and rideshares land on the right curb. Weather and Earth-observation satellites guide agriculture, wildfire response, and disaster mapping after storms and quakes. Communications constellations bring TV, telemedicine and schooling to the last mile, shrinking the digital divide. Solar panels on rooftops evolved from spacecraft power systems. Materials and power systems born for vacuum make cars safer and factories more efficient.

The list keeps growing. Precision farming with GNSS cuts fertilizer and water use. COSPASSARSAT beacons help rescuers find people in minutes, not hours. Maritime AIS and aviation ADSB relays keep ships and aircraft visible far from towers. Insurance and supply chains price risk and track goods with daily satellite imagery. 5G networks rely on space-borne timing to stay locked and low latency. Water recycling systems designed for space stations help communities survive drought. Medical monitors developed for astronauts save lives in hospitals. They are all proof that when we dream big, everyone benefits.

Space exploration isn't just for astronauts anymore. It's for the farmer thinking about sustainable agriculture, the teacher inspiring students to solve problems creatively, the nurse using technology that started in spacesuits, and the entrepreneur seeing new possibilities in recycling systems designed for lunar bases. Whether you live in a busy city or a quiet village, whether you're eight or eighty, whether you dream of walking on Mars or simply want to make life better here on Earth—space exploration belongs to you.

Looking Forward

The future isn't waiting in some distant galaxy—we are building it right now. In laboratories and classrooms, backyards and boardrooms, people everywhere are solving problems once thought impossible. We have even seen how rules and treaties will decide whether space stays open to all or becomes another place for conflict, and how responsibility—not ownership—will shape the next chapter.

We're learning to grow food more efficiently, to use energy more wisely, to work together more effectively. We're discovering that the skills needed to survive on Mars—creativity, resourcefulness, cooperation—are exactly what we need to build a better world here. This isn't science fiction. It's happening now.

And we're just getting started.

Your Invitation

We began with first principles and a promise: no complex equations, just short, stand-alone chapters you can read on a commute or between classes. From 'Science Stuff' to 'Lifting Off,' 'Orbiting Earth,' 'Exploring Space,' and the rules that keep it all fair, we have followed one thread: how ideas turn into machines—and how those machines change life here on Earth. Space is the ultimate lab for problem-solving, teaching us lessons that come back home in so many ways. Let this book be a first step in your journey to keep learning, building, and helping shape a future where the sky was never the limit—and Earth always comes out on top.

You don't need a rocket ship to be part of this story. You don't need an advanced degree or special equipment. You just need curiosity and the willingness to imagine something better. Space exploration isn't only about leaving Earth—it's about bringing back ideas that can make life better for everyone. It's about looking up at the stars and seeing not just points of light, but endless possibilities.

This is your invitation to join the journey. This frontier needs more than engineers. We need teachers to open the eyes of the next generation of talent, and students to become the new entrepreneurs. We need lawyers and policymakers to help us keep orbits safe and fair. We need welders, electricians, and construction workers to build the infrastructure. We need marketers and business people to connect innovation with customer needs. Space has never advanced because one kind of person had all the answers. The most successful space programs and companies share one trait: they pull in people who think differently, come from different backgrounds and cultures, and challenge assumptions that would otherwise go untested. Progress in space has already begun to transcend the borders of individual nation-states; the more global and diverse the team, the more resilient and imaginative the results. That is how ambitious enterprises actually win. Whether you're a student, a teacher, an artist, or an engineer—your perspective matters. Your ideas count. Your dreams can help shape the future. Wherever you stand, there's a role for you.

Space exploration isn't just about discovering new worlds. It's about building a better one right here and sharing in the responsibility to do the right thing by everyone, across countries and cultures. When more voices help design space technology, more people share in its benefits. If space has taught us anything, it is that progress is a choice—and responsibility is what turns that choice into something worth keeping.

Ad Astra—to the stars. Together.

Further Reading and Taking Action

Where Do We Go From Here?

This book captures a moment in an ongoing story, one where innovation, exploration, and adventure are always in motion.

One of the central ideas in this book is that understanding motion changes how you see the world. Once you start noticing that everything is moving, it becomes difficult to stop.

Now that you know satellites are flying around Earth at 28,000 km/h and your phone and financial markets are constantly talking to them, it is hard not to see them everywhere—on weather maps, in navigation apps, and in the systems that keep your life running. Even ordinary moments—watching the Northern Lights ripple across the sky or waiting for a kettle to boil—begin to hint at deeper stories of atoms, energy, movement, and time.

I hope this book has captured your imagination in places and offered clarity in others. More importantly, I hope your journey does not end with the final page. Like the people you have encountered here, from ancient observers to modern engineers, astronauts, and policymakers, curiosity carries you forward by asking a simple question: what might happen next?

If something in these pages caught your interest, consider it an invitation. Choose one idea, one mission, or one person and follow it a little further. Read a single book, explore an archive, study an image or dataset, or track a satellite pass and connect it back to a moment you remember from these chapters. Small, deliberate steps build understanding over time.

This book's website, everythingisflying.com, is there if you want to continue the conversation. It offers expanded resources for individuals, students, teachers, and curious readers, and it is a place to share questions, disagreements, and ideas inspired by the book. Suggestions for future editions and topics are also welcome.

What follows is a selection of books and resources that may help you dig deeper into whatever sparked your curiosity.

Early Skies and the Foundations of Understanding

The story kicks off long before rockets, with the earliest skywatchers who first began to interpret what they saw. J.L. Heilbron's *The Oxford Guide to the History of Physics and Astronomy* is a treasure trove of stories, characters, tools, and ideas that have shaped our understanding from ancient times to the space age. For those curious about personal stories of figures like Copernicus, Galileo, Kepler, and Newton, the History of Science Society's reading lists and Five Books' expert picks point to friendly, accessible histories. You can explore primary texts and modern insights on everythingisflying.com, with many editions available for free. If you're looking to understand physics more deeply without tackling dense textbooks, books like *Thermodynamics 1 With No Calculations*, *Mere Thermodynamics*, *The Simple Science of Flight*, and *Flight Without Formulae* offer clear, concept-first explanations with minimal mathematics. When it comes to fluid mechanics, which is trickier to keep completely equation-free, it's helpful to check out more visual introductions rather than traditional dense texts.

Spaceflight and the Early Space Age

For readers who want to stay longer with rockets, missions, and the first steps beyond Earth, Roger Launius's *The Smithsonian History of Space Exploration* offers a global overview from early rocketry to modern probes and commercial launch. Andrew Chaikin's *A Man on the Moon* remains the definitive narrative account of Apollo, while Tom Wolfe's *The Right Stuff* captures the culture and psychology behind the earliest days of crewed spaceflight. Robert Kurson's

Rocket Men narrows the focus to Apollo 8, revealing the risks and human decisions behind the first journey to the Moon.

For those interested in how the machines themselves were built, Roger Bilstein's *Stages to Saturn* explains the design and construction of the Saturn rockets in clear, methodical prose, while Asif Siddiqi's *Challenge to Apollo* provides a comprehensive account of the Soviet side of the space race, restoring balance to a story often told from a single national perspective.

Orbiting Earth: Satellites and Life in Space

The middle chapters of this book live largely in orbit, and Doug Millard's *Satellite: Innovation in Orbit* offers a visual and narrative history of how satellites have become an invisible but important part of our infrastructure. For readers who want to understand how orbits actually work, NASA's mission planning guides and open textbooks on orbital mechanics provide diagram-rich explanations that build intuition before mathematics.

Memoirs such as Scott Kelly's *Endurance* and Samantha Cristoforetti's *Diary of an Apprentice Astronaut* bring daily life aboard the International Space Station into sharp focus, revealing how orbital mechanics, engineering constraints, and human routines intersect hundreds of kilometers above Earth. Additional ISS research summaries and NASA e-books are linked through the companion site.

The Moon, Mars, and Further Worlds

Later chapters look outward again—toward the Moon, Mars, and the wider solar system. Maggie Aderin-Pocock's *The Book of the Moon* offers a clear, welcoming introduction to our nearest neighbor, while Oliver Morton's *The Moon: A History for the Future* explores what returning to the Moon might mean scientifically, politically, and culturally. For Mars, William Sheehan and Jim Bell's *Discovering Mars* traces centuries of observation and exploration, from telescopes to rovers. The *Outward Odyssey* series broadens the perspective further, telling the space age through human experience rather than hardware alone.

Physics and the Bigger Picture

Readers who enjoyed the physics threaded through this book may wish to widen the lens. Carl Sagan's *Cosmos* remains a model of how to weave science, history, and philosophy into a single narrative. Neil deGrasse Tyson's *Astrophysics for People in a Hurry* sketches the modern picture of the universe in compact form, while Stephen Hawking's *A Brief History of Time* and Brian Greene's *The Elegant Universe* explore deeper theoretical ideas behind gravity, time, and fundamental physics.

Law, Policy, and the Rules of Orbit

The final chapter turns toward governance and shared responsibility. The United Nations Office for Outer Space Affairs publishes the full texts of the core international space treaties, alongside plain-language explanations of how they work. Introductory texts by authors such as Frans von der Dunk, and more specialized collections in *Space Law in a Networked World*, explore how legal frameworks are adapting to commercial activity, data flows, and increasingly crowded orbits. Links to national law databases and official bibliographies are maintained through **everythingisflying.com**.

Final Word

Space has always been a collective endeavor. No one owns the sky, the rules that guide it, or the knowledge gained from exploring it. Choose a path that excites you and follow it outward.

Everything is still flying.

About the Author

Ross Hamilton is the author of *Everything Is Flying*, a book that reframes space not as distant exploration, but as the technological infrastructure shaping modern civilization. He writes for founders, engineers, policymakers, and curious readers who want to understand how complex systems actually work—and why they matter.

Trained as an aeronautical engineer, Ross began his career at Rolls-Royce Aerospace in Scotland before moving into senior technology and financial markets roles in London and New York. Over more than three decades, he has led global transformation initiatives for Fortune 100 institutions and high-growth ventures, specializing in translating deep technical capability into practical systems that scale.

He is currently the chief operating officer of Space Network, where he focuses on the commercialization of the global space sector and the integration of space technology across industries. As founder and managing partner of Sustainable Alpha, he advises startups, venture-backed companies, and investors building resilient, digitally driven businesses in space, fintech, and emerging technologies.

Ross mentors founders and leadership teams globally, helping them align innovation with long-term value. His work blends engineering discipline with commercial strategy, grounded in the belief that understanding how complex systems function is the first step toward building better ones. He believes that technological progress depends not only on engineering breakthroughs, but on the cultures, teams, and institutions that transform innovation into lasting impact.

Index

www.ingramcontent.com/pod-product-compliance
Lightning Source LLC
Chambersburg PA
CBHW020036110726
47973CB00027B/271/J